# POISONING
# THE PACIFIC

# POISONING THE PACIFIC

## The US Military's Secret Dumping of Plutonium, Chemical Weapons, and Agent Orange

## Jon Mitchell

**ROWMAN & LITTLEFIELD**

Lanham • Boulder • New York • London

Published by Rowman & Littlefield
An imprint of The Rowman & Littlefield Publishing Group, Inc.
4501 Forbes Boulevard, Suite 200, Lanham, Maryland 20706
www.rowman.com

6 Tinworth Street, London SE11 5AL, United Kingdom

Distributed by NATIONAL BOOK NETWORK

British Library Cataloguing in Publication Information Available

**Library of Congress Cataloging-in-Publication Data**

Names: Mitchell, Jon, 1974– author.
Title: Poisoning the Pacific : the US military's secret dumping of plutonium, chemical
  weapons, and Agent Orange / Jon Mitchell.
Other titles: US military's secret dumping of plutonium, chemical weapons, and Agent
  Orange
Description: Lanham, MD : Rowman & Littlefield, [2020] | Series: Asia/Pacific/
  Perspectives | Includes bibliographical references and index.
Identifiers: LCCN 2019057067 (print) | LCCN 2019057068 (ebook) | ISBN
  9781538130339 (cloth : alk. paper) | ISBN 9781538189290 (paperback) | ISBN
  9781538130346 (epub)
Subjects: LCSH: War—Environmental aspects—Pacific Area. | Nuclear weapons—
  Environmental aspects—Pacific Area. | Nuclear weapons—Environmental
  aspects—United States. | Radioactive pollution—United States. | Radioactive
  pollution—Pacific Area. | Chemical weapons disposal—United States. | Chemical
  weapons disposal—Pacific Area. | Offenses against the environment—Government
  policy—United States. | Vietnam War, 1961-1975—Environmental aspects. | World
  War, 1939-1945—Environmental aspects.
Classification: LCC TD195.W29 M58 2020 (print) | LCC TD195.W29 (ebook) | DDC
  363.7309164—dc23
LC record available at https://lccn.loc.gov/2019057067
LC ebook record available at https://lccn.loc.gov/2019057068

# CONTENTS

# AUTHOR'S NOTE

**W**henever possible, internet links to primary source documents are included in the endnotes; however, many of the reports related to Okinawa were released via the Freedom of Information Act (FOIA) and have not yet been made widely available. Key documents cited in *Poisoning the Pacific* are available on the book's homepage under the "Resources" tab at https://rowman.com/ISBN/9781538130339, with the hope that all those who have been sickened and medical professionals might be able to better understand the scope of exposure.

Also included as an appendix at the end of this book is an alphabetical list of contaminants and a brief explanation of their potential health effects adapted from the US Agency for Toxic Substances and Disease Registry.

Japanese names are given surname first; yen-dollar conversions are based on a rate of 110 yen to the dollar.

# FOREWORD

*John W. Dower*

**D**uring World War II, American song writers produced a flood of patriotic tunes addressing the war in Asia, including one titled *To Be Specific It's Our Pacific*. In the wake of Japan's defeat, a popular journalistic coinage carried this a step further by referring to the now pacified Pacific Ocean area as "the American Lake." It was taken for granted that victory did not merely establish US dominion over the Pacific and its land masses, including Japan and mid-ocean islands like Guam and Micronesia. Victory also placed the forward edge of US military might within close range of China and the Soviet Far East—no small matter in a new age of strategic air power and nuclear weapons.

In short time, the rhetoric of expansion and entitlement became framed in less colloquial terms—"national security," for instance, and consolidation of an anti-Communist "free world" governed by rule of law. "Pax Americana" became the popular label for an unprecedented global imperium of US military bases, with scores of major facilities in the former enemy nations of Japan and Germany notable among them. Today—over seven decades after the end of World War II, and three decades after the end of the Cold War—the United States maintains nearly eight hundred overseas bases, ranging in size from small garrisons to mini-cities. These are spread across more than seventy nations.

The pronounced mission of this far-flung empire of bases is, of course, to preserve the peace. And in postwar Korea and Southeast Asia, the United

States pursued peace with appalling violence. The total tonnage of bombs dropped on Korea by US air forces in the Korean War (1950–1953) was four times the tonnage dropped on over sixty Japanese cities during the American firebombing raids of 1945. In Vietnam, Laos, and Cambodia between 1965 and 1973, US forces dropped *forty* times the tonnage inflicted on Japan—and doubled up on this with massive use of toxic herbicides. In both Korea and Indochina, civilian fatalities numbered in the millions. US bases elsewhere in Asia, especially Thailand and Okinawa, provided critical support for this carnage. Okinawa's status was particularly striking, in that it was excluded from the post–World War II peace treaty that restored sovereignty to Japan in 1952, and remained under US military control until 1972. As the Harvard history professor and former US ambassador to Tokyo Edwin Reischauer noted critically in 1969, Okinawa was "a colony of one million Japanese."

Along with the Soviet Union, the great perceived enemy looming behind the wars in Korea and Southeast Asia was the People's Republic of China, which was established in 1949 after four years of post–World War II civil war. And the catchall strategic concept vis-à-vis both Cold War antagonists was "containment," a keyword dating back to 1947. The hysteria with which official Washington viewed emergence of a Communist China cannot be exaggerated. By the early 1960s, even before China tested its first nuclear weapon (in 1964), US strategic planners had identified an astounding seventy-eight Chinese cities to be targeted for a hypothetical "full force" nuclear bombing, with estimated fatalities of "107 million." (Doomsday targeting projections for the Soviet Union and its "satellite countries" were higher.) In recent years, US military publications still speak of maintaining the ability not merely to bomb China but also to penetrate the country with land forces. From where? From bases in the Pacific as well as warships with unchallenged freedom of operation in the American Lake. In American eyes, this is sound strategy. To Chinese planners, it is both provocative and humiliating. China's current aggressive activity to establish a formidable military presence in its offshore waters is a belated but predictable response to the containment policy.

All this is backdrop to the new perspective on postwar US military activity in the Pacific that Jon Mitchell presents in these pages. *Poisoning the Pacific* is a meticulous inventory of the grievous harm that nuclear tests and the deployment, storage, and disposal of hazardous materials have wreaked on land and ocean areas. But the analysis goes beyond this. The tests and bases have displaced local residents and caused illness and even deaths among them. Mitchell places this in the context of crimes against humanity and crimes against human rights, and notes that decades of callous and careless military activity have also harmed American servicemen involved in carrying out toxic activities, as well as their wives and children living on bases where contamination has taken place. We also encounter here telling snapshots of the obsequious role the Japanese government plays in its security alliance with the United States—a subservience buttressed by a boilerplate Status of Forces Agreement (SOFA) that essentially grants extraterritorial rights to America's overseas bases and military personnel. In the empire of bases, "rule of law" reinforces hierarchy, nontransparency, and nonaccountability.

The most egregious instances of environmental poisoning took place during the Cold War. In the 1950s, the United States consigned testing of its most radioactive weapon, the hydrogen bomb, to the so-called Pacific Proving Grounds. In the 1960s, US forces literally saturated Vietnam, Laos, and Cambodia with defoliants, including the now notorious Agent Orange. We already know a fair amount about this nuclear fallout and these "chemical events," some might say. But it took many years for the US government to acknowledge the long-term damage caused by its devastating practices, and even here the focus since the 1980s has been on cleaning up contamination within the United States and redressing the physical harm inflicted on its own servicemen and citizens. It took Jon Mitchell's painstaking investigative journalism to expose how Agent Orange and a host of other poisonous materials were stored and mishandled on Okinawa and elsewhere. And as documented in these pages, the cleanup of US bases in Asia has more often than not been haphazard, pollution of many kinds continues, and the United States still refuses adequate aid and redress to the thousands of Asian individuals who have suffered most severely from these crimes and abuses. The overall record that emerges, from the excesses of the Cold War to the present day, is one of obsessive secrecy, institutionalized mendacity, habitual stonewalling, thinly disguised racism, and pervasive irresponsibility.

In documenting this, *Poisoning the Pacific* draws on an impressive range of sources: written accounts in English and Japanese, extensive interviews, and many thousands of pages of formerly classified military documents accessed through the US Freedom of Information Act. As noted in passing (in chapter 9), Mitchell's investigative disclosures have been breaking news in Japan and have provoked consternation in Washington. Brought together in expanded form here, these disclosures provide not merely another incisive case study of global ecocide but also a unique window on the convoluted militarism of the so-called Pax Americana.

# INTRODUCTION

War is the ultimate environmental destroyer.

— Ui Jun[1]

**T**hroughout the twentieth century, conflicts devastated human life and the environment in ways once thought only to occur in the realm of natural disasters. Worldwide wars killed, injured, and displaced hundreds of millions. After ending, they left legacies of famine and disease; environmentally, they contaminated vast areas of the planet with lead, landmines, and unexploded ordnance (UXO). In Northern Europe, more than one hundred years after World War I, farmers continue to unearth chemical and conventional weapons, while in Laos, where fighting ceased four decades ago, 80 million American UXO still lie buried, ten explosives for every man, woman, and child in the nation. During the 1980s, the Iraqi military used hundreds of tons of mustard and nerve agents against Iranian soldiers and civilians. Then in 1991, in their retreat from Kuwait, Saddam Hussein's troops set alight oil fields, polluting land and air.[2]

Not only did such wars themselves cause untold environmental degradation, but also the manufacture of munitions exacted a fatal toll on workers and nearby communities, particularly during the Cold War, when the superpowers expended vast resources on the development of new, deadlier weapons of mass destruction. In the former USSR, military facilities leaked plumes of radioactive materials, contaminating towns, rivers, and lakes; in

1957, a nuclear dump exploded, forcing large-scale evacuations, and, in 1979, anthrax escaped from a bioweapons laboratory, exposing civilians. During the Cold War, the USSR conducted more than seven hundred nuclear tests, displacing indigenous communities and contaminating Kazakhstan and the Arctic, notably, in 1961, the detonation of the RDS-220—aka, the Tsar Bomba—which, at a strength of fifty megatons, remains the largest explosion ever produced by humankind.[3]

Other nations' militaries conducted equally reckless experiments with nuclear, biological, and chemical weapons; these tests were often held in their colonies, causing indigenous people to suffer the consequences. In the 1930s and 1940s, Britain exposed hundreds of Indian soldiers to mustard agent and, in the 1950s, detonated nuclear weapons in Australia and the Pacific, where it set safe radiation limits higher for "primitive people." France conducted almost two hundred nuclear shots in its colonies of Algeria and French Polynesia. When the environmental group Greenpeace attempted to monitor these tests in 1985, the French government dispatched saboteurs to blow up their ship, the *Rainbow Warrior*. China, too, detonated dozens of nuclear weapons in Lop Nur, exposing hundreds of thousands of Muslim Uighurs to radioactive fallout.[4]

During the last eighty years, no nation's military has damaged the planet more than that of the United States. Since 1941, the United States has been at almost constant war, causing extreme environmental contamination. Its atomic bombings of Japan, while taking two hundred thousand lives in 1945, produced widespread radioactive fallout. In Vietnam, Laos, and Cambodia, its use of Agent Orange and other herbicides sickened millions of Southeast Asians, US troops, and their allies; so severe was this environmental destruction that it gave birth to a new word: *ecocide*. More recently, US-led wars in the Middle East have contaminated communities with lead and depleted uranium, while military burn pits in Iraq and Afghanistan have released untold toxins, sickening local civilians and US troops alike.

Between 1945 and 1992, the United States exploded more than eleven hundred nuclear bombs, mostly in the Pacific region and the deserts of the United States. These tests dispossessed local peoples, poisoned their health, and contaminated their lands, in some cases, so badly that they may never be able to return. In addition, hundreds of thousands of US troops and civilians—today known as "downwinders"—were exposed to fallout. As well as detonating full-scale weapons, US scientists tested the effects of radiation in smaller, more pernicious ways. During the Cold War, in the name of national security, they allowed reactors to melt down in outdoor stress tests, vented radioactive particles to judge their dispersal in the environment, and conducted experiments on prisoners, mentally disabled children, and expectant mothers.[5]

Accompanying these radiation tests, the US military also ran trials of chemical and biological weapons on unsuspecting populations. In the 1950s and 1960s, it spread bacteria over San Francisco and in the New York subway to simulate bioweapon attacks and, until the mid-1970s under the auspices of Project 112, exposed service members to toxins, including nerve and mustard agents, without their knowledge.[6]

Just as in the Soviet Union, the factories that manufactured the Pentagon's Cold War weapons of mass destruction created severe pollution. In 1979, for example, a dam breeched at Church Rock uranium mill in New Mexico, contaminating Navajo people's water supplies with radioactive materials; likewise, the facilities that built the atom bombs dropped on Japan, including Hanford Site in Washington State and Los Alamos National Laboratory in New Mexico, were contaminated, as were the factories that produced Agent Orange for the Vietnam War. The very weapons designed to protect Americans and their allies have ended up poisoning them instead.

For more than seven decades, the Western Pacific region—Japan, Okinawa, and Micronesia—has borne the brunt of US military contamination. Hundreds of thousands of civilians, service members, and their families have been exposed to toxic substances, including radioactive fallout, nerve agents, and dioxin; water, air, and soil have been polluted, and some areas have been rendered uninhabitable for the foreseeable future.

During World War II, US troops fought throughout the Pacific to liberate colonized people from brutal Japanese control, but following the war, instead of allowing their subjugated populations to enjoy their new freedom, the United States turned their islands and territories into military-administered colonies and used them as test grounds for weapons of mass destruction. In the Marshall Islands, the United States detonated dozens of nuclear devices, and on Okinawa, it conducted biological agent tests and packed the island with nuclear warheads and chemical munitions.

Since the 1950s, the military has used its facilities in the region to wage conflicts around the world, leaving their indelible mark on both the war zones and its Pacific bases in the form of contamination from Agent Orange, depleted uranium, and an alphabet soup of such poisons as trichloroethylene (TCE), polychlorinated biphenyls (PCBs), and per- and polyfluoroalkyl substances (PFAS). Today, the peoples of the Pacific continue to live with the consequences of past and present military operations. In the Marshall Islands, radioactive fallout from US nuclear tests prevents the return of inhabitants, while contaminated sites abound on Guam, Saipan, and Tinian; on Okinawa, pollution from military operations impedes the redevelopment of former base land, and, as later chapters show, the United States has laced the drinking water of one third of the island's population with carcinogenic fire-fighting chemicals.

For as long as the military has been contaminating the Pacific, it has also tried to hide its damage by covering up accidents and downplaying their impact; however, there are two ways to cut through this smokescreen: First, via the US Freedom of Information Act (FOIA) and, second, by interviewing former and current military personnel, base workers, and whistleblowers. As an investigative journalist, these are my two prime tools, and this book is the culmination of a decade of research, bringing together more than ten thousand pages of documents from the US military, State Department, and Central Intelligence Agency. Dating from World War II to the present, these reports reveal the vast extent of contamination and the US military's attempts to conceal it; key documents have been uploaded to the publisher's homepage. Additionally, this book includes interviews with Americans, Japanese, Okinawans, and Micronesians who possess detailed knowledge of the workings of the US military—people who are so concerned about safety conditions that they feel unable to stay silent.

During World War II, many countries researched chemical and biological weapons, but only Japan used them in combat—a violation of international agreements. Chapter 1 explores the small island of Okunoshima, where Japan produced toxic agents and deployed them in China, killing thousands. Following surrender, the weapons were dumped at sea and buried throughout Japan and China, where they continue to injure people. Japan also had an industrial-scale biological weapons program in Manchuria led by Unit 731 scientists, who conducted tests on Chinese prisoners and unleashed typhoid, cholera, and plague on the civilian population; it planned attacks on the United States, too. Following surrender, US officials granted immunity to suspected Japanese war criminals in exchange for their research data, which they then used to develop their own weapons of mass destruction.

Chapter 2 explores the impact of US nuclear weapons in the Pacific. After the atomic bombings of Hiroshima and Nagasaki, the United States concealed the spread of radiation and censored reports in the media. During the Cold War, the United States conducted sixty-seven nuclear tests in the Marshall Islands, displacing islanders, particularly at Bikini Atoll, and polluting large areas with radioactive fallout, forcing local people to evacuate to avoid contamination, which still lingers today. In 1954, the CASTLE Bravo thermonuclear explosion—one of the world's worst environmental catastrophes—sickened Marshallese and Japanese fishermen; thousands of US soldiers were again exposed to radiation during cleanup operations in the late 1970s.

Chapter 3 starts with the Battle of Okinawa, which sowed the small island with UXO, while taking the lives of more than one-quarter of the civilian population. In the twenty-seven years following the end of World War II, the United States turned Okinawa into a military colony that its troops nicknamed the "Junk Heap of the Pacific." Bases possessed one of the largest arsenals of weapons of mass destruction on the planet, and accidents involving these munitions sickened service members and civilians, and pose an ongoing risk today.

During the Vietnam War, the US military sprayed millions of liters of experimental herbicides, principally the most infamous, Agent Orange, across Indochina. Chapter 4 examines how the United States has wrapped its herbicide program in lies—concealing the chemicals' usage, their toxicity, and the locations where they were used. Hundreds of veterans insist that they stored, sprayed, and dumped Agent Orange on Okinawa, the most important staging point for the Vietnam War, but the government denies that the chemicals were ever present there and refuses to help most of the sick service members.

Chapter 5 examines how the US military became the worst polluter on the planet. Throughout the twentieth century, its wars contaminated foreign countries with dioxin, depleted uranium, and UXO; in the United States, forty thousand sites have seeped poisons into the ground and local water sources. In recent years, the federal government has forced its military to become more environmentally responsible in the United States, but in Japan, the US military's seventy-eight facilities are still allowed to contaminate with impunity according to the terms of the Japan–US Status of Forces Agreement (SOFA). As a result, time and time again, pollution has been discovered on returned base land, causing grave health and economic consequences.

Chapter 6 lifts the lid on how bases have contaminated present-day Okinawa, host to thirty-one facilities and home to fifty thousand Americans related to the military. FOIA-released reports and interviews with whistle-blowers reveal damage from depleted uranium, solvents, and PCBs. Unlike in the United States, the military has refused to take responsibility for the contamination, and the Japanese government can't—or *won't*—do anything to intervene.

Chapter 7 explores how Japan, the only nation to suffer wartime atomic attack, came to embrace nuclear power. Due to a propaganda campaign run by US intelligence officials and Japanese conservatives, more than fifty nuclear plants were built in Japan, ignoring the risks of earthquakes and tsunami. After the 2011 meltdowns at the Fukushima Daiichi nuclear power plant, US service members were exposed to radiation while operating in the Tohoku region, and, unbeknownst to the Japanese public, the military dumped radioactive water into sewers beneath its bases.

Today, the United States possesses sixteen territories unprotected by the full extent of the constitution. Chapter 8 explores how the military has contaminated Guam, the Commonwealth of the Northern Mariana Islands, and Johnston Atoll. Guam was exposed to fallout from the first US full thermonuclear blast in 1952, but residents have received no compensation. Meanwhile, CIA operations on Saipan littered the island with abandoned hazardous materials. Johnston Atoll has suffered from the full spectrum of US weapons of mass destruction, polluted with plutonium from failed nuclear launches, used as a dumpsite for Agent Orange, and finally transformed into a factory to decommission chemical weapons.

The US military is the most powerful organization in the world, and the Pacific communities it has poisoned are some of the poorest. Despite their seeming powerlessness, many people in the region refuse to let the military continue to contaminate their land. Chapter 9 profiles some of the activists and organizations demanding environmental justice and proposes fairer guidelines that prioritize human health and the environment to protect the region from further damage in the decades to come.

*Nuchi du takara.*
Life is precious.

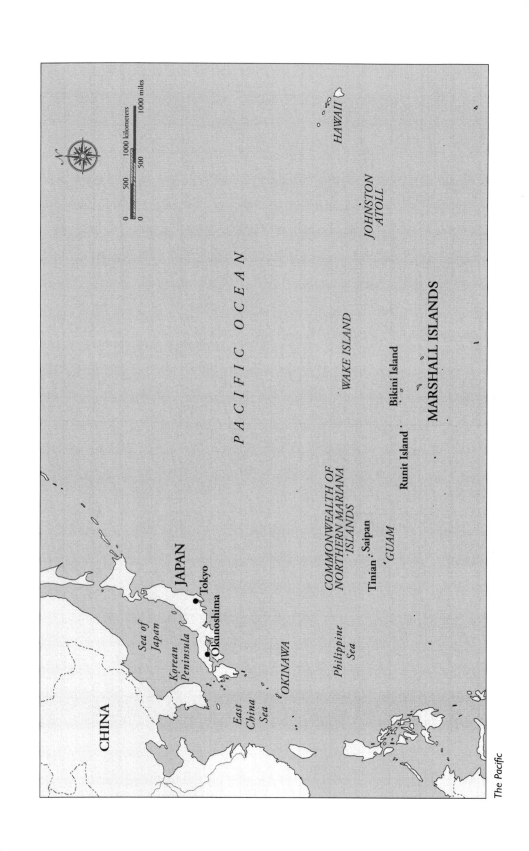

The Pacific

# 1

## JAPANESE WEAPONS
## OF MASS DESTRUCTION
## AND THE US COVER-UP

Relations between the United States and Japan date to 1853, when US gunboats sailed into Tokyo Bay and demanded the nation open to trade. For centuries, Japan had closed itself off from the world, threatening those who left or landed with execution, according to a policy known as *sakoku*, which had insulated it from the colonization suffered by other Asian countries but, at the same time, left it lacking the industrial advances that had become commonplace in the West. The arrival of US warships and the subsequent imposition of unequal trade treaties convinced Japanese leaders they needed to modernize—and *fast*—to avoid further exploitation. In the following decades, under the Meiji Reform, Japan introduced sweeping reforms, central to which was the creation of a vast national army.

As well as adopting such Western innovations as steam power and industrial production, Japan also embraced its love of colonialism. For a half-millennium, European nations had been exploiting Asia for its natural resources, and now, by the end of the nineteenth century, Japan was hungry to build its own empire. In 1895, it annexed Taiwan, and ten years later it defeated Russia, becoming the first modern Asian country to beat a European power; in 1910, Japan occupied Korea. With the outbreak of World War I, Japan saw an opportunity to expand its empire, and so, based on an alliance with Britain, it declared war on Germany and seized its colonies in China and Micronesia.

At postwar Versailles, sitting down on the winners' side of the table, Japanese leaders proposed a clause guaranteeing racial equality in the hope they would receive the international recognition as an imperial power they had long been seeking—but it was rejected. The refusal cemented Japan's suspicions that the coming years would set it on a collision course with Western powers in Asia, and as a result it ratcheted up the modernization of its military and, what was de rigueur at the time, the creation of a state-of-the-art chemical weapons program.

During World War I, approximately 135,000 tons of chemical agents had been used on the battlefields of Europe, killing 90,000 soldiers and injuring 1 million. Although Japan's troops had not participated in this combat, its military leaders were aware of the munitions' effectiveness and keen to harness their power. In 1919, Japan established the Army Institute of Scientific Research in Tokyo to study chemical weapons, and in the following years, Japanese scientists traveled to France, Germany, and the United States to learn production techniques. In 1923, Japanese research suffered a setback when a 7.9-magnitude earthquake struck Tokyo, destroying many of the military's laboratories and forcing it to seek a new site.[1]

## OKUNOSHIMA: POISON GAS ISLAND

The army chose the small island of Okunoshima, Hiroshima Prefecture. Located three kilometers from the mainland in the Inland Sea, it was isolated enough to maintain secrecy and prevent widespread loss of life in case of accident; at the same time, Okunoshima was close to key ports and rail lines so its products could easily be transported domestically and overseas.

In 1927, the army evicted Okunoshima's seven households and placed the island under the Military Secrets Act, making it off limits to the public. In the following years, the government constructed a 150-building complex of chemical weapon production facilities, including factories, laboratories, test centers, and warehouses. Buildings were painted green and brown to camouflage them from the air, while trains traveling along the mainland coast were ordered to lower their blinds when passing the area; in 1932, the authorities removed the island from maps.[2]

On Okunoshima, the army produced five chemical weapons: tear gas, vomiting agent, hydrogen cyanide, lewisite, and mustard agent. Vomiting agent was comprised of the arsenic compounds diphenylcyanarsine (DA) and diphenylchloroarsine (DC); packed into small flares, cannisters, and barrel-sized tubs, when lit by a fuse, the smoke caused sneezing, vomiting, and temporary blindness; inhaled at high concentrations, it was lethal.[3]

Lewisite had first been manufactured by US scientists in the final weeks of World War I, too late to deploy in combat. With a distinctive smell of geraniums, it caused blisters, blindness. and bloody noses; it could also trigger "lewisite shock," whereby damaged capillaries harm the kidneys and lead to fatally low blood pressure.

The weapon for which Okunoshima was best known was mustard agent. Named after its pungent odor, the chemical's short-term effects included watery blisters and blindness. Long term, mustard agent has been linked to lung and respiratory cancer, and damage to DNA. Unlike such gaseous chemical weapons as chlorine, mustard agent is viscous; after deployment, it can remain in the soil, staying dangerous for weeks. On Okunoshima, scientists mixed mustard agent with lewisite in a formula that stayed a liquid at low temperatures, enabling its use during the subzero winters of Northern China, which Japan had occupied since 1932.[4]

The Japanese military assigned a color code to each of its chemical weapons. Red Agent was the vomit-inducing DA and DN; Yellow Agent was a mix of mustard and lewisite. Quality control was assured at a test laboratory where rabbits were gassed in glass cases or shaved and exposed to blister agents.

Manufacturing Okunoshima's chemical weapons was a workforce of sixty-six hundred adults, often keeping the facility running twenty-four hours a day. For the most dangerous work, they wore full-body rubber suits hooked to ventilators, which workers nicknamed octopus suits due to their goggles and proliferation of tubes. But the suits sprang leaks, and Yellow Agent caused severe skin and respiratory problems. Although a clinic on the island provided rudimentary care for sick workers, its doctors were more concerned about sending records of patients' symptoms to military research centers, where the data was used to develop more potent weapons.[5]

With the start of the war against the United States, more Japanese adults were conscripted to fight overseas, creating labor shortages at home. Elsewhere these were filled with forced workers from China and Korea; however, the secrecy of Okunoshima's mission restricted it to Japanese nationals. Starting in June 1943, more than one thousand children between the ages of thirteen and fifteen were brought to work on the island, where they assembled smoke cannisters and sewed ten-meter-wide paper balloons designed to drop biological weapons and explosives on the United States. Disposing of the waste materials and stacking barrels exposed these children to the same toxins as adult laborers. No area of Okunoshima was untouched by contamination. Red pine trees, once the pride of island residents, were defoliated by fumes so strong they also corroded windows; toxic runoff killed marine life.[6]

As well as Okunoshima, there were at least thirty-four other production and storage depots for chemical weapons in Japan. At Kokura City, Fukuoka Prefecture, one thousand workers, again including children, produced Blue Agent, phosgene. Ten times more toxic than chlorine and with a smell of fresh hay, upon inhalation it reacts with moisture in the lungs to form hydrochloric acid; it is often fatal. At Kokura, too, workers fell ill in large numbers.[7]

Sagami Naval Factory, Kanagawa Prefecture, was the site of the Japanese navy's chemical weapons production facilities. Opened in 1943, it manufactured approximately 760 tons of poison agents during its short run. These munitions were designed to support guerilla operations against the upcoming Allied invasion of the homeland. Bottles labeled "Beer" were filled with mustard agent and lewisite, ready to lob at enemy troops, while, throughout the nation, women and children trained to fight with bamboo spears.[8]

During World War II, the major powers produced chemical weapons, but Japan was the only country to use them in combat.

## JAPANESE CHEMICAL WEAPONS IN COMBAT

Since the start of its occupation of Manchuria in 1932, Japanese rule had been characterized by brutality, corruption, and exploitation of the local population. Japanese officials saw Manchuria as a place where they could conduct activities impermissible elsewhere. In 1937, the Japanese military launched an invasion of the rest of China; outnumbered by Chinese forces, Japan sought any way to gain an advantage. From the outset of the conflict, it deployed chemical and biological weapons.

In Wuhan, Hubei Province, for example, between August and November 1938, Japanese troops used approximately thirty-two thousand grenades and ten thousand artillery shells of Red Agent. Army manuals described the results of such attacks: Enemy troops writhed, choking on the floor, and Japanese soldiers, accidentally exposed by a change in wind direction, "evacuated their bowels." For some Chinese troops, Red Agent was lethal: "When the soldiers came in contact with special smoke, some seriously affected persons bled through their noses and mouths and died from asphyxiation."[9]

During raids, Japanese troops employed small quantities of Red Agent as a knockdown gas to incapacitate their victims. In one case in March 1944, in Gaoyang County, Hebei Province, six Chinese men captured in this manner were then taken to Japan and forced to work in Nagano and Gifu prefectures.[10]

As the war in China escalated, Japanese troops deployed the full range of Japan's Okunoshima arsenal. In October 1941, for example, in Ichang, the Japanese army fired more than three hundred bombs containing mustard agent and the blood agent hydrogen cyanide, resulting in six hundred fatalities. Then, in February 1942, Japanese soldiers spread approximately three hundred tons of mustard agent in Taihan, Shanxi Province, killing hundreds.[11]

The Japanese military's use of chemical weapons violated international agreements. Japan was a member of the 1907 Hague Convention, and it had signed—but not ratified—the 1925 Geneva Agreement, explicitly outlawing the use of chemical and bacteriological weapons. To conceal their actions on the battlefield, Japanese military commanders ordered their troops to retrieve as many chemical barrels and cannisters as possible. But Japan's use of toxic munitions was so widespread that it was impossible to hide, and word soon reached the international community. In May 1938, the League of Nations passed a resolution condemning Japanese chemical attacks in China, and the US press carried reports that, by 1942, there had been at least one thousand mustard and lewisite attacks against Chinese troops.[12]

In June 1942, President Franklin D. Roosevelt condemned Japanese use of chemical weapons as an "inhuman form of warfare" and warned that if such use persisted in China and elsewhere, "retaliation in kind and in full measure will be meted out." The warning deterred the Japanese military from the organized use of chemical weapons against white troops. There were only a handful of uses in Guadalcanal, involving toxic smoke, and on Okinawa, although chemical weapons were discovered after the war, there were no recorded instances of their actual use in combat.[13]

However, in China, Roosevelt's warning went unheeded, and by the end of the war, Japan had used chemical weapons an estimated two thousand times, resulting in as many as forty thousand military and civilian deaths.[14]

As with its deployment of biological weapons, after the war, no Japanese leaders were ever prosecuted by the Allies for their nation's use of chemical weapons. One exception occurred in Hong Kong in 1948, when the Australian military prosecuted two Japanese soldiers for exposing an Australian POW to hydrogen cyanide in Indonesia in 1944. Japanese troops had wanted to check whether their chemical weapons had exceeded their shelf life so they threw them at a captured Australian air force captain and a Dutch East Indies air force sergeant. The weapons were still potent—and, in postwar Hong Kong, the Japanese soldiers were sentenced to death.[15]

During the war, the United States, too, had drawn up plans to use its own stockpiles of chemical and biological agents. Major Japanese cities would be drenched in phosgene and mustard agent; rice fields would be targeted with anticrop chemicals. Meanwhile, an arms factory in Indiana was tasked with producing tens of thousands of anthrax bombs to spread disease throughout Japan. Ultimately, however, the United States decided to deploy another weapon of mass destruction instead.[16]

## THE ALLIES ARRIVE ON OKUNOSHIMA

Between August 15, 1945, when Emperor Hirohito announced his nation's surrender, and the arrival of Allied troops in Japan two weeks later, the nation's skies turned black as officials burned documents that might be used as evidence against them in war crimes trials. On Okunoshima, too, the army incinerated reports, dismantled some production facilities, and dumped munitions at sea; workers burned their employment papers, fearing American reprisal.

Wartime intelligence had alerted the Allies to the existence of Japan's chemical weapons program so American officers arriving in Tokyo conducted interrogations to discover its location. In October, Allied troops landed on Okunoshima, where they found the factories largely intact and vast stockpiles of chemical weapons.

At the time, Australian troops administered the region, but with little experience in handling chemical weapons, they asked the United States for assistance. Experts from the army's Chemical Warfare Service arrived on Okunoshima and drew up plans to remove the poisons. With memories of dangerous conditions fresh in their minds, the island's previous workers refused to return, so the Allies ordered the Teikoku Rayon Company to send a crew. In the following weeks, 300 of these workers gathered approximately 15,000 tons of weapons and 3,000 tons of base chemicals from Okunoshima and neighboring islands.[17]

The US military devised a two-stage plan: first destroy the weapons, then destroy the facilities. Starting in May 1946, workers loaded the poisons onto two American ships and ran a pipe from the island's Yellow Agent storage tanks to pump the poisons onto the vessels. One man died when he was

*Following Japanese surrender, Allied forces ordered civilians to gather barrels of chemical weapons and bombs on a beach on Okunoshima; this August 1946 photograph shows the ordnance before it was loaded onto ships, which were then sunk in the Pacific.* Australian War Memorial

exposed to blister agent on one of the ships, and another ninety were injured when the hose from shore to ship was severed in a storm. Following the accident, an American officer ordered workers at gunpoint to retrieve the broken pipe from the seabed; one man developed blisters from head to toe, hospitalizing him for three months. After the loading work had been completed, the two ships sailed south, where, overlooked by an Allied submarine, they were sunk in the Pacific.[18]

Back on Okunoshima, the Allies faced the problem of how to dispose of the remaining stocks of Red Agent, which due to its buoyancy could not be dumped at sea. Workers packed some of the weapons into air-raid shelters and burned them; one worker lost his arm when the propellant exploded. Tens of thousands of vomit agent cannisters were stacked in caves and doused with sea water and bleach, a mixture the Allies believed would render safe the arsenic-based munitions. The caves' entrances were then plugged with concrete blocks.

Stage two of disposal operations began in May 1947. Allied officers ordered laborers to tear down Okunoshima's main buildings and raze the ruins with flamethrowers, work that enveloped the island in dense smoke. Today the walls of the remaining buildings are still blackened, and the island's interior is littered with pipes and chunks of the ceramic pots once used to manufacture blister agent.

*On Okunoshima in March 1947, Japanese workers, following Allied orders, used flamethrowers to incinerate the interior of chemical weapon factories; the smoke enveloped the island, and many workers fell sick.* Australian War Memorial

The 1947 cleanup operation was so fatiguing that the deputy director of one of the Teikoku Rayon factories died from overwork; however, there was a payoff for the company—the United States allowed it to take large stocks of chemicals to manufacture insecticides. In the early 1950s, the US military renovated some of Okunoshima's buildings into arsenals for its own weapons used in the Korean War. Local communities suspect US chemical munitions were stored on Okunoshima in the same way they were brought to Okinawa at this time.[19]

The cleanup of Okunoshima failed to remove the dangers posed by the island's chemical weapons. Fishing ships, for example, hauled up dumped ordnance. In May 1958, a twenty-kilogram hydrogen cyanide bomb netted near Okunoshima was taken to a scrapyard, where it leaked, killing one and injuring twenty-seven. On the island itself, two workers were exposed to suspected Yellow Agent in July 1955, after finding contaminated clothing in a pond; one of the men died from his injuries.[20]

In the 1960s, against the advice of former employees who knew the dangers of their previous worksite, Japanese authorities decided to transform Okunoshima into a national vacation village. Japanese Self-Defense Forces (JSDF) were tasked with conducting surveys for remaining chemical weapons. In 1961, they discovered in an air-raid shelter enough cannisters of Red Agent to fill five or six trucks; then, again in 1970, government officials discovered hundreds of cannisters of smoke and vomiting gas in a cave, but instead of removing them, they merely sealed them in place.[21]

In 1995, government surveys at several locations on Okunoshima detected serious arsenic contamination from Red Agent in soil, and the groundwater, the source of the island's drinking water, was also polluted. Seeking a safe supply, authorities decided to run a pipeline from the mainland to Okunoshima, but they had to abandon the plan due to concerns the construction work might dislodge chemical weapons dumped on the sea bed. Today, all water consumed on the island is delivered by weekly boat.

Nine out of ten workers involved in Okunoshima's wartime production and postwar cleanup suffered serious health damage, especially respiratory tract cancers and chronic bronchitis; others committed suicide. Initially, the Japanese government avoided helping the survivors. In 1961, following demands from former full-time workers, it finally began to award them monthly financial aid; in 1965, it offered payments to the families of the deceased. Support for part-time workers, child laborers, and those involved in the cleanup operations took even longer to be approved, a situation complicated by some claimants having destroyed their own work papers due to fears of Allied prosecution. In 1975, the Japanese government started providing free medical treatment and a small monthly allowance—but only after victims met strict screening criteria.[22]

The Japanese authorities' response left many workers feeling betrayed. They had been poisoned while supporting the nation's war effort, but the government then did all it could to evade responsibility. The Ministry of Education has kept from most textbooks the military's use of chemical weapons in China so today few Japanese people are aware of this history.

Now, Okunoshima is a popular tourist destination, nicknamed "Rabbit Island," thanks to hundreds of the animals that, according to who you ask, either escaped from the island's World War II laboratories or, more likely, were brought there to attract visitors in recent years. The rabbits burrow alongside the air-raid dumpsites and hump among the piles of broken pots once containing blister agents. A poison gas museum was opened in 1988, but it is often bypassed by visitors more attracted to the Instagram charm of the bunnies.

Local guide and chemical weapons expert Yamauchi Masayuki says the Japanese government would prefer that people forget what happened on Okunoshima. "Thousands of chemical weapons remain buried here. The ground is still contaminated, and the water can't be drunk. There is the danger of contamination throughout the island, but tests have only been conducted for arsenic and none for other substances such as lead and mercury."[23]

No one knows how contaminated the rabbits are.

Japanese chemical weapon production also impacted the communities that produced raw materials for Okunoshima's factories. In Shimane Prefecture, a production area for the ingredients of lewisite, rice fields were poisoned and people fell ill. Beginning in 1937, there was an outbreak of serious illnesses in another area producing raw materials for chemical weapons, this time in Fukuoka Prefecture, sickening twelve thousand and killing more than seven hundred. At the time, it was blamed on dysentery, a theory the chief of the local water board strongly rejected. Twenty-five years later, researchers discovered that, prior to the outbreak, there had been an explosion at the factory, which, they believe, contaminated the water; the company denied the charge.[24]

Manufacturers of the raw chemicals, which reaped large profits from military contracts, have paid no compensation to survivors. Reincarnated under new names, their businesses boomed during the postwar years, especially during the Korean War, when the United States pumped billions of dollars into Japanese companies to supply material for its troops.

## THE POISONED HOMELAND

In 1998, high levels of arsenic were detected in wells providing drinking water to Kamisu Town, Ibaraki Prefecture. As many as five hundred residents fell ill, and authorities suspected contamination from Red Agent was to blame. The incident prompted the Japanese government to launch a nationwide investigation into postwar dumps of chemical ordnance.[25]

Published in 2003, the survey gathered the testimonies of Japanese veterans and reports from the US and Australian militaries. The results stunned the Japanese public, most of whom either knew little of their nation's wartime chemical weapons or assumed they'd been stored in only a handful of arsenals. The survey revealed that poison munitions had been dispersed throughout Japan during the war for testing; stockpiled for an Allied invasion; and then, as air raids intensified, moved to areas deemed less of a target. Following surrender, Japanese and Allied troops had dumped these caches at more than forty sites. The area worst affected was the Pacific Ocean off the coast of Choushi City, Chiba Prefecture, where US forces had sunk at least 450 tons of chemical weapons in popular fishing grounds, a decision that would later haunt residents, as more than six hundred incidents involving these munitions would occur in the coming decades.[26]

The US military had been irresponsible in its initial sea dumps—but the Japanese government's delay in reacting to the problem was equally negligent. In March 1970, it finally initiated cleanup operations in the Choushi area, but only eleven chemical weapons were retrieved, and eight more people were injured. That same year, another cleanup lasting from mid-September to early October was more successful, clearing about sixty items, with no recorded injuries.[27]

In recent years, the number of reported discoveries has dropped significantly; however, this may be due to fears of *fuuhyou higai* (harmful rumors). Fishing crews may indeed have been netting ordnance but not informing the authorities to avoid negative media coverage, which might damage sales of locally caught seafood.

Fishing crews have been injured in other parts of Japan, including Tokyo Bay and coastal areas of Kyushu. In July 1985, for example, in Hitachi City, Ibaraki Prefecture, several children suffered injuries to their eyes and skin while playing next to a metal tube netted by fishermen and left on the wharf side.[28]

There have also been high-profile incidents inland. In 1995, in Dejima, Hiroshima Prefecture, widespread arsenic contamination of the soil was discovered. At the end of World War II, a private company had bought surplus materials from the production of Red Agent and abandoned them; the prefecture buried them in 1973. The official response drew public anger; delays were made in the cleanup, and the national government left prefectural authorities to pick up the tab, despite the chemicals' origins under military contract.[29]

In 1996, in the northernmost prefecture of Hokkaido, Japanese authorities discovered twenty-six 50-kilogram blister agent bombs at the bottom of Lake Kussharo, a source of local drinking water. Following their removal, they were buried in nearby woods and, four years later, destroyed by international experts.[30]

Kanagawa Prefecture, home to Japanese naval production facilities, has also experienced numerous incidents. Between 2002 and 2003, more than a dozen construction workers were injured by such abandoned chemical weapons as mustard agent, lewisite, and hydrogen cyanide. Authorities subsequently unearthed more than eight hundred bottles of guerilla warfare munitions and discovered widespread soil contamination.[31]

## CHINA: ABANDONED WEAPONS, ABANDONED PEOPLE

China has suffered the most from abandoned Japanese chemical weapons. During World War II, Japan shipped approximately 7.5 million shells there, and following surrender, munitions that had not been used in combat were either left in place or dumped in lakes and rivers; almost all documents related to their original deployment, use, and disposal were destroyed. Since the war, these abandoned chemical weapons have been discovered at more than ninety sites, and approximately two thousand Chinese people have been injured.[32]

For decades, the Japanese government refused to admit it had brought chemical weapons to China—let alone do anything to clear them away. Then, in 1992, the Chinese government presented data at a conference in Geneva, claiming the Japanese military had abandoned some two million chemical weapons in their country; later it reduced its estimate to four hundred thousand. The announcement came in the run-up to negotiations concerning the Chemical Weapons Convention, a treaty that sought to eliminate stockpiles of such weapons, including abandoned ones. Realizing it would have to clean up its dumped ordnance, Japan was initially reluctant to sign the agreement; however, the United States pressured Japan into joining to compel China to destroy its own stockpile of modern chemical weaponry.

In 1993, Japan signed the convention; in 1999, it finally admitted that it had abandoned chemical weapons in China and agreed to clean them up. Since then, safety, logistical, and political problems have delayed any real progress. The rusty weapons are leaking toxins and, in many places, are mixed with conventional explosives, increasing the risks for cleanup teams. The locations of the dumps are often difficult to access and, in Northern China, frozen ground limits the time when the work can be conducted. Final cleanup costs have been estimated at $9 billion—to be paid by Japan—but, in 2007, remediation had to be halted due to a financial scandal involving the Japanese company contracted to perform the work.[33]

Some of the largest hurdles have been political. In Japan, conservatives still deny its military deployed chemical weapons to China, and neo-nationalist sound trucks often picket exhibitions attempting to educate the public on the problem. Subsequently, there is little popular support, let alone awareness, of the cleanup attempts. On the Chinese side, too, the government has exploited the issue to score points in Sino–Japan relations, while, on the ground, Japanese contractors have been held to higher environmental standards than their local counterparts.

At times, these factors have slowed the pace of remediation to a stand-still. It was not until September 2010, that the first chemical weapons were destroyed, and as of March 2018, only forty-nine thousand weapons, 12 percent of the estimated total, had been eliminated.[34]

The legacy of Okunoshima's chemical weapons is not unique to Japan and China; it is a common one for militaries all around the world. Nations spend massive amounts of time, money, and scientific resources manufacturing the most technologically advanced weapons possible; however, they always omit one important point: how to dismantle the munitions if they end up abandoned or unused. The problem is especially urgent when it involves weapons of mass destruction—the United States has struggled for decades to demilitarize its own arsenals of chemical ordnance. In many cases, at the time of the production of such weapons, the technology does not even exist to dismantle them and must be developed from scratch. The hurdles are not only technological, but also economic. Swept up in their grand schemes, war planners neglect to factor in the costs of cleanup; there is always enough money to build new weapons but rarely enough to render them safe.

## JAPANESE BIOLOGICAL WEAPONS

Compared to the chemical weapons that previously had been produced by most major powers before, during, and after World War I, research into biological weapons was a slower process. Britain did not begin until 1940, and the United States held off until 1943, before setting up a research center at Camp Detrick, Maryland. With a starting date of 1932, Japan was ahead of the pack, motivated, as with chemical weapons, by a desire to exert control in Manchuria, where its troops were outnumbered, and defend the long, isolated border with the Soviet Union.

Heading Japanese research on biological weapons was Ishii Shiro, a doctor educated at prestigious Kyoto University, who led the army's Unit 731. Working alongside Ishii were hundreds of other scientists based at more than twenty sites in the Japanese empire, two of which were Singapore and Burma.[35]

In 1932, the Japanese military began to research biological weapons at Zhong Ma Prison Camp in Beiyinhe, Heilongjiang Province. Scientists exposed human subjects, including anti-Japanese guerillas, Russian prisoners of war, and criminals, to three main diseases: anthrax, glanders, and plague.[36]

Following a prison break in 1934, the military closed Zhong Ma Prison Camp and relocated its headquarters for biological weapons research to a much larger site at Ping Fan, measuring six square kilometers. After forcibly removing six villages, the military built a complex tailored to the mass production of biological weapons: greenhouses to test anticrop diseases, barns to experiment on animals, a runway from which Unit 731's air wing could conduct bomb trials, and prisons for human subjects. Japanese officials told residents the complex was a lumber mill, and from this pretense sprung the code by which Japanese scientists called their human subjects: *maruta* (logs). At least three thousand *maruta* were killed at Ping Fan, and many others died at sites across the empire.[37]

Conducting research at Ping Fan were approximately one thousand scientists, many of whom, like Ishii, were graduates of Japan's top universities. To keep them enjoying a lifestyle to which they were accustomed, the Ping Fang complex was equipped with a swimming pool, bars, and a geisha house; middle and high schools were also built for their children. Throughout the late 1930s and early 1940s, scientists shuttled back and forth between Ping Fan and universities in Japan, publishing papers, giving lectures, and recruiting graduates for the biological weapons program.[38]

## *MARUTA*

At Ping Fan, scientists expanded their research from the three diseases studied at Zhong Ma Prison Camp to other human pathogens with the potential for weaponization, including cholera, typhoid, and gas gangrene. Production of these diseases took place on a massive scale. Pathogens were grown on thousands of jelly-like culture plates; at its peak these could produce hundreds of kilograms of bacteria a month. Fleas were bred in large metal containers holding rats and rice husks; Ping Fan had an estimated four thousand such containers monthly producing millions of fleas.[39]

From a military perspective, one of the largest challenges of biological weapons is delivering them to their target—explosives, for example, incinerate the pathogens before they can infect their victims. To overcome this problem, Japanese scientists experimented with bombs made from glass and ceramics that would fragment into shards of infected shrapnel. In tests, these bombs were loaded with fleas and dropped onto targets surrounded with flypapers; scientists then counted the number of fleas stuck at each distance from the bomb. Moreover, Japanese scientists fitted aircraft with equipment to spray biological weapons from low altitudes and infected hundreds of horses and sheep with anthrax spores.[40]

Human experiments involved tying *maruta* to stakes at fixed distances from bombs to investigate how they could be infected by such diseases as anthrax and gas gangrene. Prisoners were also given milk, chocolate, and melons infected with typhus and bubonic plague. Following exposure, test subjects were dissected while still alive and without anesthetics, which doctors feared would muddy their data.[41]

In Manchuria, Japanese researchers also conducted human experiments with the chemical weapons produced at Okunoshima. For example, in outdoor tests in 1940, hundreds of shells of mustard agent were fired into the midst of Chinese prisoners to ascertain the effectiveness of different kinds of shelters and protective gear. Other *maruta* were forced to drink Yellow Agent, and it was dripped into prisoners' eyes. Ping Fan also possessed a chamber where human subjects, including at least one mother and child, were gassed with hydrogen cyanide. While the Japanese were conducting such experiments in China, on the other side of the world, the Nazis were using hydrogen cyanide in the form of Zyklon-B to kill millions of Jews, gypsies, and political enemies.[42]

Leaving no doubt that knowledge of Unit 731 tests permeated the highest levels of Japanese society, the Emperor's younger brother, Prince Mikasa, recalled in his memoirs watching films of Chinese prisoners "made to march on the plains of Manchuria for poison-gas experiments on humans."[43]

Outside China, Japanese scientists conducted research in Singapore, breeding rats and fleas carrying bubonic plague; in the Philippines, prisoners of war were used in experimental amputations and dissections. Meanwhile, at Kyushu Imperial University in May and June 1945, medical teams—unaffiliated with Unit 731—subjected eight American POWs to live experiments, including removal of their lungs, injections with seawater, and drilling into their brains.[44]

## BIOLOGICAL WEAPONS IN COMBAT

The extent to which the Japanese military used biological weapons in combat is still widely debated. Following the war, for example, Ishii claimed there only had been twelve such instances, but today most experts believe that number is far higher.

One of the earliest deployments of biological weapons occurred along the Soviet–Manchuria border in 1939. According to former Unit 731 members,

Japanese troops poured typhoid and other bacteria into the Khalkin-Gol River, hoping it would sicken the Russian troops camped downstream. The results were inconclusive, but Ishii's unit received a commendation following the action suggesting his superiors were satisfied with the results.[45] Throughout 1939 and 1940, the Japanese military continued to conduct field tests with typhoid, poisoning approximately one thousand wells in the Harbin area.[46]

The following October, there were two well-documented attacks on cities in Chekiang province, where Japanese aircraft scattered a mixture of grain and fleas over Ningpo and Chuhsien. According to the *Chinese Medical Journal*, wheat had likely been dropped to attract the local rat population, which would then become infected by the fleas. Bubonic plague later spread in the areas, resulting in a reported 120 deaths; the disease had not occurred in the region before. A similar attack took place in 1941, in Changde, where thirty-six kilograms of plague-carrying fleas were dropped, sparking an epidemic that killed 7,643 residents.[47]

In April 1942, the US military launched its first-ever air attack on Japan. Known as the Doolittle Raid, eighteen US bombers took off from ships in the Pacific Ocean, struck the Tokyo area, and then landed in China, where the crews were helped by civilians and troops. Both the raid and local assistance infuriated the Japanese military, which took revenge by massacring a quarter-million Chinese. As well as conventional weapons, the Japanese army employed biological munitions, contaminating wells and reservoirs with cholera, typhoid, and plague. Some of these attacks backfired, infecting Japanese troops, seventeen hundred of whom died.[48]

In 1942, cholera epidemics blamed on Japanese biological weapons spread in Yunnan and Shandong provinces, killing four hundred thousand people. Elsewhere, Japanese airplanes dropped feathers riddled with anthrax, and troops left pens and canes contaminated with plague, and doled out rolls spiked with typhoid. Many of these operations were carried out to provide a veneer of legitimacy for Japan's scorched-earth policies—when an epidemic broke out, the military would burn down the village on the pretext of disease control.[49]

Against this backdrop of warfare, mass displacement, and starvation, it is difficult to accurately pinpoint the number of biological weapons victims. US historian Sheldon H. Harris estimates the total in the "six-figure range,"[50] and Chinese experts have put the number at 580,000. On the other hand, Japan's leading Unit 731 researcher and no defender of its actions, Tsuneishi Keiichi, has argued there is no evidence of widespread epidemics sparked by Japanese biological weapons use.[51]

Also controversial is whether victims of Japanese bioweapon tests included Allies. Without a doubt, Russian prisoners were tested, and there was the incident where eight POWs were dissected at Kyushu University, but other American POWs believe they were used in human trials in the Mukden POW camp. According to the US government, no conclusive evidence has been found.

On the other hand, the Japanese military did explore at least three ways to use biological weapons against American troops and civilians. In 1944, during the Battle of Saipan, Japanese military scientists traveled to the island with a cargo of plague-carrying fleas to distribute around its runway—but their ship was sunk by US submarines before it reached its destination. The second plan involved arming the paper balloons produced by children on Okunoshima and elsewhere to carry biological weapons on the Pacific gulf stream to the United States. According to Japanese scientists, many tests were conducted, but they eventually decided balloons were "unsatisfactory for this purpose." Instead, the Japanese military loaded the balloons with conventional explosives and sent six thousand of them aloft. Approximately two hundred reached the United States, killing at least seven people. US authorities censored news of the attacks to avoid creating panic. The final bioweapons' plan, codenamed "Cherry Blossoms at Night," involved Japanese airplanes dropping plague-infected fleas over San Diego. Scheduled to start in September 1945, the war ended before it could be put into action.[52]

## THE US COVER-UP

After Japan's surrender, its military in China attempted to erase all traces of its biological weapons program. At Ping Fan, the remaining *maruta* were gassed with hydrogen cyanide, documents were incinerated, and the buildings were demolished; local laborers were killed so they could not reveal what had happened there. Most perniciously, Japanese scientists released the test animals, including horses infected with glanders and rats carrying bubonic plague. In the following years, these animals were linked to epidemics in the region.[53]

Ahead of the 1.5 million Soviet troops advancing through Manchuria, Japanese scientists fled to Japan. At home, they attempted to blend into anonymity and avoid detection by Allied authorities; Ishii was so scared of arrest that he requested a newspaper run an article stating that he had been killed and asked his friends to hold a mock funeral for him.

Occupation authorities arriving in Japan were aware of the biological weapons program. From Chinese sources, the United States had received information about plague outbreaks, and its own interviews of captured Japanese troops had provided details of Unit 731's research. US forces tracked down key members of Unit 731, one of whom was Ishii, and both sides began a drawn-out game of bluff and counterbluff: The Americans wanted human test data and wanted to keep it out of Soviet hands; Japanese scientists were equally wary of the Soviets and wanted immunity from the upcoming war crimes prosecutions in Tokyo.

Between 1945 and 1947, experts from Camp Detrick conducted interviews with Japanese scientists, and although the full extent of the information given to the United States may never be known, a report from June 1947 gives a sense of its scope. It summarizes how Japanese scientists wrote for the Americans a sixty-page paper on "BW activities directed at man" and provided information, including maps and tactics, related to twelve field tests against Chinese civilians and soldiers. Another nineteen-page report detailed biological weapons tests on crops using "fungi, bacteria, and nemotodes"; the Camp Detrick author concluded it contained "much interesting and worthwhile information." The Americans also interviewed other Japanese scientists specializing in biological weapons for use against animals.[54]

As the interviews progressed, Japanese scientists admitted to still possessing eight thousand slides detailing how biological weapons had affected some two hundred human cases. "These had been concealed in temples and buried in the mountains of southern Japan," the scientists told the Americans.[55]

In July 1947, letters from the assistant chief of staff of General Headquarters revealed the military's delight at obtaining the human test data, describing it as the "only information available in world" and the field tests as the "only known tests in actual warfare." The military concluded, "Information procured will have the greatest value in future development of the US BW program." In return for the data, Japanese scientists had received "food, miscellaneous gift items, and entertainment," plus direct payments totaling between 150,000 and 200,000 yen. According to the military, this was a "mere pittance."[56]

Five months later, another report from Camp Detrick was equally as effusive, stating, "Such information could not be obtained in our own laboratories because of scruples attached to human experimentation." The author went on to recommend, "It is hoped that individuals who voluntarily contributed this information will be spared embarrassment because of it and that every effort will be taken to prevent this information from falling into other hands."[57]

To spare from embarrassment the Japanese scientists who had committed countless crimes against humanity, the United States granted them full immunity from prosecution. In the Tokyo Trials, 5,700 Japanese war criminals and collaborators were charged, but no Japanese were punished for the use of biological weapons. US prosecutors also kept the issue of chemical weapons out of the court for fear it would lead to questions about biological weapons and boomerang into accusations its atomic bombings of Hiroshima and Nagasaki were a form of poison warfare.

Decades later, when the immunity agreement became public, one of the original Tokyo Trial judges, B.V.A. Röling, was livid, stating, "It is a bitter experience for me to be informed now that centrally ordered Japanese war criminality of the most disgusting kind was kept secret from the court by the US government."[58]

In 1949, the Soviet Union held its own trial of Japanese scientists, sentencing twelve to prison. At the time, General Douglas MacArthur dismissed the prosecutions as Communist propaganda. Meanwhile, the Soviets were also taking advantage of Japanese biological weapons research, with thirty Japanese scientists working at a factory near Moscow.[59] Both the United States and the USSR had rehabilitated German experts in a similar manner, putting them to work on their rocket and atomic programs.

In the following years, rumors abounded about the postwar activities of former Unit 731 scientists. One of them was linked to the sensational 1948 robbery of the Teikoku Bank in Tokyo, where twelve people were poisoned by a man claiming to be a public health officer treating an outbreak of dysentery in the neighborhood; others were linked to a secret US laboratory at Yokosuka Naval Base, Kanagawa Prefecture. The most persistent accounts place Japanese former military scientists collaborating with the United States to launch biological weapons attacks during the Korean War. The alleged assaults involved US airplanes dropping disease-infected fleas and anthrax-laden feathers; captured US airmen admitted to the raids but later withdrew the confessions, claiming they'd been made under duress. A report by an international scientific commission concluded the bioweapons assaults appeared to be "developments of those applied by the Japanese army during the Second World War." The CIA dismissed the reports' authors, who came from the United Kingdom, Sweden, and the Soviet Union, as "commie-liners."[60]

Such rumors have never been categorically proven or debunked, and in some ways, they detract from the abhorrent—and well-documented—reality of what Unit 731 scientists did after receiving immunity from prosecution. Many of them enjoyed successful careers in Japan, often parlayed on their human tests. They joined the faculty of top universities, entered pharmaceutical companies, and worked for the national medical association; one of them became governor of Tokyo and another the head of the Olympics Committee. Many former Unit 731 scientists found work at the US Atomic Bomb Casualty Commission, taking tissue samples from Hiroshima and Nagasaki survivors—often under duress—and sending the data to the United States to improve nuclear warfare. Three of the former Unit 731 top scientists went on to establish Japan's largest medical company, Green Cross, which hit the headlines in the 1980s, when it infected more than two thousand hemophiliacs in Japan, South Korea, and the United States with HIV from untreated blood.[61]

As for Ishii himself, in addition to gifts and money from the United States, he received a large retirement pension from the Japanese government; he opened a clinic in Tokyo and died from throat cancer in 1959.

As with Japanese chemical weapons, Chinese people suffered the most from germ warfare. Because most documents had been destroyed or locked away in the United States, for more than a half-century the Japanese government was able to deny its troops had waged biological warfare in China. No reparations were paid to Chinese victims, and it was not until the early 1980s that the Japanese government even acknowledged the existence of Unit 731.

In 1997, 180 Chinese victims and relatives of those killed by Japanese biological weapons filed a lawsuit in Tokyo seeking compensation from the Japanese government. During the next five years, twenty-eight hearings were conducted, in which Chinese witnesses and Unit 731 veterans testified about spreading biological weapons through the civilian population. The plaintiffs demanded compensation for twenty-one hundred victims whose identities and deaths could be confirmed.

In 2002, the Tokyo court finally reached its verdict—and the ruling was unprecedented. Judge Iwata Koji concluded the Japanese military had violated both the Hague and Geneva conventions. "The evidence shows that the Japanese troops, including Unit 731, used bacteriological weapons under the order of the Imperial Japanese Army's headquarters and that many local residents died."[62]

The ruling contradicted decades of official denials and ought to have paved the way for justice for Unit 731's victims. But instead, the court stated they were not entitled to any compensation because according to peace treaties made with Japan in the 1970s, China had relinquished all such rights.

The immunity deal brokered between US occupation forces and Japanese army scientists was one of the most significant in the postwar period. The pact signaled to Japanese militarists that if they collaborated with the United States, their past misdeeds—no matter how evil—would be not only forgiven, but also rewarded. It set the stage for further collusion between the United States and suspected Japanese war criminals in the coming years as Washington built Japan into a bulwark against Communism. The United States turned its back on the Chinese victims of Japanese biological and chemical weapons despite China having been a US ally during World War II and the slaughter of the more than 250,000 people who rescued its Doolittle pilots. In the coming Cold War, the United States ruled these victims better left forgotten, an inconvenient truth in the new world order. US occupation authorities and Japanese politicians deterred public debate on the matter, blanketing it in a wall of secrecy that, in Japan, lasted until the 1980s; even today it is rarely discussed and still denied by Japanese conservatives.

Just as importantly for the United States, its deal with Unit 731 scientists and other suspected war criminals also ensured the silence of postwar Japanese leaders regarding its own use of weapons of mass destruction in World War II. The United States tacitly agreed to let Japan abandon its Chinese victims and, in return, prevented Japan from criticizing its atomic bombings and the hundreds of thousands suffering from radiation poisoning in Hiroshima and Nagasaki. In this way, from the very beginning, US–Japan postwar relations were built on a foundation of violence and complicity that absolved their military's leaders of guilt and disregarded the rights of civilians.

# 2

# NUCLEAR WARFARE IN JAPAN AND THE MARSHALL ISLANDS

**D**uring the first decades of the twentieth century, scientists around the world were racing to split the atom in the hope it would unleash tremendous amounts of energy that could be harnessed for the production of nuclear power or the creation of a bomb thousands of times more destructive than conventional explosives. The 1930s saw a series of leaps in nuclear physics, culminating in 1939, when fission was first sparked in uranium; by the early 1940s, with the world at war, the major powers were trying to militarize the atom but experienced little success. Germany's efforts were hampered by poor coordination between different research groups and Allied raids on production facilities, whereas Japan was stymied by shortage of funds and a lack of dedication among scientists, who doubted the efficacy of a nuclear bomb; instead, Japan focused on the manufacture of chemical and biological weapons.[1]

The United States was the only nation with the financial resources and a large enough pool of scientists, many having fled Nazi persecution in Europe, to pursue production of an atomic bomb. Backed by $2 billion from the government and the mobilization of some of the richest corporations in the country, within three years the Manhattan Project became "as large as the entire automobile industry of the United States at that date," using as much as 10 percent of the nation's entire electric supply and employing 125,000 workers. While the rest of the world lacked the resources to build even a single type of atomic bomb, the United States simultaneously pursued two: one loaded with enriched uranium and the other with pluto-nium, a potentially more destructive element first discovered in 1940.[2]

Producing the world's first nuclear bombs was hazardous work. At the University of Chicago in 1942, scientists built a prototype reactor beneath the stands of the American football pitch, putting the campus—if not the entire city—at risk in case of an accident. Later, at the plutonium process-ing plant at Hanford, Washington, the military decided to speed up produc-tion of plutonium by cutting its cooling time, a shortcut that endangered workers' safety. Today, both Hanford and Oak Ridge, Tennessee, where uranium was enriched, are still contaminated from Manhattan Project work and postwar bomb production. Meanwhile, at the main research facility at Los Alamos, New Mexico, scientists brought together half-spheres of plutonium to estimate their criticality in a process they called "tickling the dragon's tail." Following the war, the experiment killed two researchers who mishandled the grapefruit-sized core.[3]

Scientists lacked knowledge about the health damages of radiation. Although they understood X-rays were dangerous—in 1934, for example, radiation researcher Marie Curie had died from leukemia—they knew nothing of the impact of the full-body exposure that would occur following a nuclear explosion. Manhattan Project scientists were even less aware of the spread of fallout following a blast. Some team members raised concerns about such effects, but no serious studies were undertaken, even as they edged closer to completing a working device.

On July 16, 1945, the US military was ready to test the world's first nu-clear bomb—"Gadget"—at the Trinity Test Site in New Mexico. Uncertain of its success, some scientists gave it only a slim chance of working, but, on the other hand, in case they had underestimated the explosion, they warned the New Mexico governor to be ready to declare a state of emergency.

At 5:29 a.m., Gadget exploded with an estimated strength of twenty-one thousand tons of explosives, four times larger than expected. Its flash burned through photographic film and temporarily blinded witnesses fourteen kilometers away. The fireball's heat fused the surface of the desert to glass, and the blast wave felled one witness at nine kilometers. Fallout burned the hides of cattle sixty-four kilometers to the north, and several farming families, according to the US Department of Energy, received "significant exposures."[4]

The director of the test, Kenneth Bainbridge, called the explosion a "foul and awesome display." Robert Oppenheimer, director of the Los Alamos laboratory, recalled a line from the Bhagavad Gita: "Now I am become Death, the destroyer of worlds."[5] The explosion rattled nearby communities, forcing the military to put out a press release explaining there had been an accident at an ammunition depot—the first of many cover-ups in this new nuclear age.

## URBAN TEST SITES: HIROSHIMA AND NAGASAKI

Within a month of the Trinity shot, the military conducted its second and third tests—on urban targets. In the summer of 1945, Hiroshima was home to 350,000 residents and, like most Japanese cities at the time, embedded with military facilities, notably a port that shipped Okunoshima's chemical weapons to China and a new army headquarters built in anticipation of the Allied invasion. While military infrastructure took up a fraction of Hiroshima, the rest of the city was civilian. Thus far, the United States had deliberately spared Hiroshima from conventional air raids to better assess the impact of the new bomb. In the words of Colonel Paul Tibbets, who piloted Hiroshima's destruction, Hiroshima and Nagasaki were "good virgin targets" perfect for "bomb damage studies," and President Harry S. Truman himself referred to the bombing of Hiroshima as an experiment.[6]

On August 6, the US military tested its second nuclear device. Loaded with approximately 60 kilograms of enriched uranium, "Little Boy" exploded 600 meters above downtown Hiroshima, destroying or severely damaging everything in a 5-kilometer radius and shattering windows almost 20 kilometers away.

Three days later, a B-29 carried the third nuclear bomb built by the United States to Japan. The primary target was Kokura City, where Allied intelligence suspected the presence of the Japanese army's phosgene factory, but clouds and smoke obscured the city, making the crews divert to Plan B. With a population of 263,000, Nagasaki, too, held military infrastructure, particularly the munitions plant that had built the torpedoes fired at Pearl Harbor, but it was primarily civilian; like Hiroshima it, too, had been excluded from prior attacks to make it easier to survey the bomb's impact.

The plutonium-packed "Fat Man" detonated three kilometers off target, about five hundred meters above Urakami district, the largest Catholic community in the country. The US Army's official follow-up report stated the explosion had "even greater blast effects than in Hiroshima. Total destruction spread over an area of about three square miles (eight square kilometers). Over a third of the fifty thousand buildings in the target area of Nagasaki were destroyed or seriously damaged."[7]

Splitting the atom over the skies of Japan instantaneously released energy equivalent to thousands of tons of high explosives—approximately fifteen thousand tons at Hiroshima and twenty-one thousand tons at Nagasaki. Japanese *hibakusha* described the fission as a *pika*—the flash—followed by a *don*—the blast. Near the hypocenters, the *pika* vaporized thousands, transforming humans to shadows, stains on concrete, and charcoal mannequins. Farther from the hypocenter, the flash ignited hair, charred skin, and seared holes through clothes; dark colors absorbed rays more easily, leaving patterns burned into their wearers' skin. Scientists at Trinity and the B-29 bomb crews were equipped with welders' glass or sunshades to protect their eyes; those in Hiroshima and Nagasaki lost their sight to the flash. "A piece of Japanese paper exposed nearly 1.5 miles (2.4 kilometers) from X (the hypocenter) had the characters which were written in black ink neatly burned out," laconically noted the army's follow-up report.[8] In Hiroshima, the flash seared people as far as 2.3 kilometers from the hypocenter, and at Nagasaki, people were burned 4.2 kilometers away. Temperatures on the ground reached four thousand degrees Celsius.

Moments after the *pika* flash came the blast wave of the *don*. Traveling at speeds of three kilometers a second, it stripped off clothes and hair; survivors were left raw as their skin, already charred by the flash, was blown from their bodies. The blast demolished wooden buildings, shattered concrete, and bent steel; windows broke at such velocity that their shards scarred stone. Trapped beneath the wreckage, survivors burned alive.

Seconds after the blasts, air rushed in to fill the vacuum, sucking ash, soot, and dust into a column that, in the case of Nagasaki, reached twenty kilometers high. Only an estimated 10 percent of the bombs' uranium and plutonium cores reached fission, so the remainder—approximately sixty kilograms of highly radioactive material—dispersed in the mushroom-shaped clouds.[9]

In both cities, radioactive plumes drifted far and mixed with cooler air to create rain. In Hiroshima, this rain—blackened by ash—fell heavily over an area of two hundred square kilometers, while lighter rain fell over the rest of the city; clothes doused on that day still contain traces of radioactive cesium-137. Following the explosions, rescue teams and relatives flocked into the devastated cities to help survivors and search for loved ones; they ate contaminated food, drank contaminated water, and breathed contaminated dust.[10]

Soon people in the bombed cities began to fall sick from radiation poisoning. In the first few days, they suffered nausea, diarrhea, and bloody vomit; in the following weeks, they lost hair, bled from their gums, and

*On August 10, 1945, the day after the detonation of the world's third atomic bomb, an elderly woman crawls through debris twelve hundred meters from the hypocenter in Nagasaki. Photograph by Yamahata Yosuke. Courtesy of Yamahata Shogo*

developed purple spots on their skin. Radiation damage made survivors bleed and weakened their production of red and white cells. Thousands, including those who'd escaped visible injury from the initial blasts, died from reduced bone marrow function, intestinal bleeding, and infectious diseases their bodies were unable to resist. Babies exposed in the womb were born dead or with microcephaly and extreme mental disabilities.[11]

Because the explosions destroyed public records, calculating a precise death toll from the blasts is difficult. Estimates for the number of fatalities by late 1945 are 140,000 at Hiroshima and 80,000 at Nagasaki. Among the dead were thousands of non-Japanese: Southeast Asian students, Koreans, and Chinese, some of whom

*A young girl stands next to a charred corpse three hundred meters from Fat Man's hypocenter in Nagasaki on August 10, 1945. Photograph by Yamahata Yosuke. Courtesy of Yamahata Shogo*

were forced laborers. At Hiroshima, military casualties were approximately 3,240—2.3 percent of the total dead—and 250 at Nagasaki—0.3 percent of the total. American military personnel died as well—twenty-three POWs in Hiroshima, some lynched by Japanese in retaliation for the bombing, and eight POWs in the Nagasaki blast.[12]

In World War II, the indiscriminate bombing of civilians was nothing new. Dresden had been destroyed by Allied bombers and Chinese cities targeted by Japanese conventional, chemical, and biological weapons. By the time the United States struck Hiroshima and Nagasaki, most Japanese cities had also been leveled by US bombers dropping napalm, phosphorus, and magnesium munitions; however, what made the atomic bombs different was the ongoing suffering that lasted long after they had been detonated. In the following years, *hibakusha* developed cataracts, and those with flash burns that had initially seemed to heal developed thick keloid scars that left them in agony. Five to ten years after the explosions, *hibakusha* began to develop cancers of the colon, lungs, and stomach. Children started to fall ill with leukemia seven to eight years following the blasts; the younger the age at which they were exposed, the earlier it developed. Child leukemia rates reached eighteen times the national average, and the overall cancer risk of *hibakusha* was 40 to 50 percent higher than normal. In Nagasaki, even the city's vegetation was impacted, with mutated flora observed for several years after the bomb had been dropped.[13]

As early as August 10, Japanese scientists determined that their country had come under atomic attack, but the government ordered them to hide their findings, and most Japanese people did not realize what had happened until the emperor's surrender announcement of August 15. Soon after the blasts, Japanese journalists and filmmakers documented the damage. Then, less than a month after the bombings, the first foreign reporters arrived. Australian Wilfred Burchett described Hiroshima as looking as if it had been flattened by a "monster steamroller"; American George Weller described Nagasaki as "frizzled like a baked apple."[14]

Both journalists witnessed *hibakusha* dying in droves from radiation exposure, which Weller referred to as "Disease X." Realizing the illnesses were unprecedented, the two journalists filed reports about the atomic bombs' health effects. On September 5, Burchett's article was published in the British *Daily Express* under the headline, "The Atomic Plague: I write this as a warning to the world."

The US military's reaction was swift. Two days later, it convened a press conference in Tokyo, where General Thomas Farrell, the Manhattan Project's head officer in the Pacific, explained there had been no possibility of residual radioactivity and Burchett "had fallen victim to Japanese propaganda."[15] Meanwhile, from Nagasaki, Weller sent reams of articles to Tokyo to be relayed to his editor at *Chicago Daily News*, but almost none of them saw the light of day for sixty years. The military censor, he later recalled, disposed of them in the "upper drawer of the 'file and forget.' . . . Twenty thousand skulls pulverized in an hour beside Nagasaki's dour creek—who would believe them censorable today?"[16] Sadly, such censorship set the standard for the US military to follow for the next seventy-five years.

## CENSORSHIP AND SUPPRESSION

America's disinformation campaign about its atomic assault on Japan started within hours of its bombing of Hiroshima. On August 6, when Truman announced the attack, he called the city an "important Japanese army base." Four days later, he doubled down on the falsehood by claiming Hiroshima had been selected to avoid civilian casualties.[17]

US authorities were particularly sensitive about accusations that the atomic bomb was a form of poison warfare. In the weeks immediately following the bombings, they dismissed or downplayed the impact of radiation, claiming the number of radiation deaths was "very small." In late 1945, the director of the Manhattan Project, General Leslie Groves, went as far as to proclaim that radiation poisoning was a "very pleasant way to die."[18]

The Manhattan Project's follow-up report denied any damage from fallout: "Radiation from scattered fission products and induced radioactivity from objects near the center of explosion were definitely proved not to have caused any casualties."[19]

Such false assurances ought not be blamed on only military malfeasance, but also the timing of US radiation checks. Nuclear detonations create thousands of radionuclides, some of which would have decomposed by the time US scientists arrived on the scene. Moreover, both cities experienced storms in the weeks after the bombings that washed away radioactive particles from the barren terrain into rivers and the sea.

Regardless of the reasons, early US assertions were wrong. Modern estimates place the number of radiation-related deaths as high as 20 percent of the total, and both the US and Japanese governments recognize that residual radiation caused serious health damage. Today, Japan categorizes as *hibakusha* those who entered the cities within two weeks of the bombings, and the US Department of Veterans Affairs awards assistance to

POWs and members of the Occupation forces present in either city from August 6, 1945 to July 1, 1946.[20]

During the US occupation of Japan, the military introduced censorship that silenced arguments against its official disinformation campaign regarding the attacks on Hiroshima and Nagasaki. Starting with a broad warning from President Truman on September 14, to the US media not to report details of the bombings, several days later, Occupation authorities in Japan introduced the Press Code, requiring Japanese media to be screened before publication.[21]

More than eighty-five hundred members of the Occupation's Civil Censorship Detachment set about removing references to issues deemed critical of the Occupation, including preparations for war crime trials, food shortages, and the existence of censorship itself. When it came to the atomic attacks, the United States censored mention of fatalities, survivors' testimonies, and photographs from newspapers, textbooks, and private letters on the grounds of "disturbing public tranquility," "causing resentment to Allied Powers," and "incitement to unrest."[22]

In September 1945, Occupation authorities suspended two Japanese newspapers for criticizing the dropping of the bombs, and they commandeered from Japanese and American camera crews footage of the aftermath to ship to the United States, where it was kept secret for decades. Also sent to the United States were the medical reports made by Japanese doctors who'd been treating *hibakusha*.

In 1946, the United States created the Atomic Bomb Casualty Commission (ABCC) to monitor survivors in Hiroshima and Nagasaki. The commission offered no medical treatment; instead, its staff only took photographs, as well as blood and tissue samples, which were sent to the United States for use in nuclear war planning, including, many suspect, the design of more powerful weapons.

Working alongside the American doctors at the ABCC were former Unit 731 scientists dispatched by the Japanese government at the request of Occupation authorities.

For these men, the work of the ABCC was a seamless extension of their human experiments in China. Alongside American doctors, they harassed *hibakusha*, threatening with prosecution those who refused to cooperate in their research and pressuring bereaved families to allow loved ones' bodies to be autopsied and sent to the United States. Commenting on Japanese doctors' work at the ABCC, former Unit 731 vivisectionist Kojima Saburo called it a "golden opportunity" to research radiation victims.[23]

The *hibakusha's* belief that they had been abandoned by their own government was cemented by the 1952 Treaty of San Francisco, which absolved the United States of all reparations, thus preventing *hibakusha* from seeking compensation in the years to come. The Japanese government had sacrificed its own citizens to cozy up to the United States, a pattern that would play out time and time again in the following decades. It was not until 1957 that the Japanese government eventually began offering limited medical support to *hibakusha*, and it took until 1978, for foreign victims, including the estimated thirty thousand Korean *hibakusha*, to become eligible for help—but even then they faced such problems as the requirement to produce a Japanese witness when applying for support.[24]

By the time US censorship ended in 1949, it had significantly added to the suffering of *hibakusha*. Japanese doctors had been unable to share information with colleagues about how to treat victims, no medical symposia addressed radiation exposure until 1951, and censorship hampered *hibakusha* struggles to receive welfare support. Censorship caused ignorance about radiation sicknesses, encouraging discrimination against *hibakusha* due to fears their illnesses might be contagious; *hibakusha* themselves were unable to understand what was happening inside their own bodies. The silencing of *hibakusha* voices and suppression of photographs of their injuries reduced public awareness of nuclear warfare to images of mushroom clouds and desolated landscapes—awful but devoid of any *human* suffering.

US authorities filled the void left by their censorship with one of the most successful propaganda campaigns in history: the argument that the atomic bombings of Hiroshima and Nagasaki had ended World War II. In February 1947, *Harper's Magazine* published US Secretary of War Henry Stimson's article "The Decision to Use the Atomic Bomb," which argued that the dropping of the bombs ended World War II, preventing 1 million US casualties.[25]

There are many arguments against the assertion. Truman's own papers suggest the president thought the bombs were worse than chemical-biological weapons and had not actually been needed to end the war. The opinion was supported by top US military commanders, who, later in public or private, stated the war was already over by the time the bombs had been dropped. That the bombings targeted the Soviets as much as Japan is almost certain; in the words of Secretary of State James F. Byrnes, the United States dropped the bombs to stop the Soviets "get so much in on the kill" and prevent their occupation of Japan, and the postwar carve-up of former occupied territories.[26]

On the other hand, support for the argument that the bombs ended the war came from an unexpected quarter: Emperor Hirohito. In his surrender address, he stated,

> The enemy has begun to employ a new and most cruel bomb, the power of which to do damage is, indeed, incalculable, taking the toll of many innocent lives. Should we continue to fight, not only would it result in an ultimate collapse and obliteration of the Japanese nation, but also it would lead to the total extinction of human civilization.[27]

Already on the verge of surrender, the bombings provided the Japanese government with a moral pretense to end the war: It was acting to save the whole of humanity. Under this guise, Japan's defeat was not due to any failings of its politicians, military, or industrialists, or the Emperor himself—instead the bombs gave them a convenient deus ex machina to end the war without losing face and allowed for many to return to powerful positions in the coming years without any punishment or serious reflection on their brutal, imperialist policies.

Today, Henry Stimson's argument remains prevalent in the United States, and to question it still invites accusations of anti-Americanism and historic revisionism. In 1994, when the Smithsonian Museum decided to hold an exhibition exploring the decision to drop the bombs from a variety of perspectives, the backlash from conservative groups forced it to abandon the plan.

The US cover-up of its nuclear bombings of Japan not only muted public criticism, but also allowed the United States to pursue postwar nuclear research with minimal oversight. At home, US scientists conducted nuclear experiments on US citizens, releasing radiation from the Hanford Site in the notorious Green Run test in 1949, irradiating prisoners' testicles, and feeding radioactive substances to pregnant women. Its rocket program was abetted by Nazi scientists whose work raining V-2 rockets on Britain was forgiven, as they were granted amnesty under Operation Paperclip similar to Japanese Unit 731 scientists.

In the United States, nuclear weapons tests exposed more than two hundred thousand troops to fallout, but secrecy agreements made them maintain silence for decades. Civilians downwind from the detonations were contaminated, too; among those worst affected were Native American communities living in the deserts near the test sites. In the Pacific Ocean, US nuclear tests were even more reckless.[28]

## THE MARSHALL ISLANDS

Following the surrender of Japan, many Manhattan Project scientists wanted a halt to the nuclear weapons program; however, the US government decided to expand it instead. Nonetheless, the larger-than-expected blast at Trinity and the apocalyptic damage in Japan made them nervous about testing such devices in the United States, so they sought a site overseas.

The Marshall Islands are located four thousand kilometers from Hawaii, and, in 1945, their inhabitants were no strangers to colonial oppression. First occupied by Germany between 1885 and 1914, their islands were then seized by Japan, which built up military infrastructure there, in violation of a mandate by the League of Nations. Japanese troops forced residents to work on these projects and seized civilian food. Even on tiny Bikini Atoll, Japan installed a military outpost, accompanied by a ban on Christian services, and forced labor on islanders.

In January 1944, the United States launched its attack on the Marshall Islands at Kwajalein. The battle lasted a week, killing approximately 370 US and 8,000 Japanese troops, plus an estimated 200 Marshallese. When the fighting stopped, said one US soldier, the island looked as if "it had been picked up to twenty thousand feet and then dropped."[29]

Following the war, the Marshall Islands became part of the Trust Territory of the Pacific Islands, controlled by the United Nations (UN). Soon, however, the UN granted the United States strategic trusteeship of the territory, with the understanding that Washington would promote economic and social advances; the UN's ultimate goal was for the United States to encourage the islands to become self-governing and independent.

But the military regarded the region as war booty—the blood it had shed to win the islands had earned it the right to treat them as they wanted, even if it meant destroying them. For the US government, seeking overseas sites for their atomic tests, the Marshall Islands ticked all the right boxes. Their isolation allowed secrecy, and scientists hoped the Pacific Ocean would absorb radiation with minimal harm. Moreover, their population was relatively small; as Henry Kissinger said in later years, "There are only ninety thousand people out there. Who gives a damn?"[30] With their minds made up, the only question was how to break the news to the residents of Bikini Atoll.

On Sunday, February 10, 1946, the US military governor of the Marshall Islands met the islanders as they were leaving church and told them their home had been chosen for a special mission. Likening them to the "Children of Israel whom the lord had saved from their enemy and led into the Promised Land," he explained that US scientists had devised a weapon and were planning to "learn how to use it for the good of mankind and to end all world wars."[31]

Prior to the governor's arrival, the islanders generally had a favorable view of the United States; it had liberated them from the Japanese and developed some infrastructure. Hearing the governor's explanation, the islanders expected the relocation to be temporary, and at least some of them feared the repercussions if they tried to refuse.

One month later, the military transported Bikini's residents to the atoll of Rongerik. Here, the land was smaller and less fertile, and the fish in the lagoon were poisonous; they soon began to starve. In the following years, the Bikinians were shuffled from island to island as the military expanded nuclear tests throughout the region. Residents of Enewetak Atoll were treated in a similar way—relocated by Americans, who understood nothing about their dietary or cultural needs, to islands where self-sufficiency was impossible.

## THE GUINEA PIG FLEET

In the summer of 1946, the United States assembled what it called a "guinea pig fleet" in Bikini's lagoon. The ninety-five military ships, two of which had been captured from Japan and one from Germany, were anchored into rings and spokes, resembling a massive target; at its bullseye floated the USS *Nevada*, painted bright orange to make it easy to spot from the air. The ships' decks were loaded with more than fifty-five hundred rats, pigs, and goats; US Army biological weapons experts also brought cases of live bacteria to the test to judge how radiation might affect them.[32]

Operation CROSSROADS Able took place on July 1, 1946. A B-29 bomber dropped the world's fourth atomic bomb, a plutonium device identical to the one used on Nagasaki. Exploding with a similar twenty-one-kiloton yield, it sunk five ships and severely damaged six others; all the vessels were contaminated by radiation. About 20 percent of the animals died in the blast, with half dying later from burns and radiation poisoning.[33]

From start to finish, the Able test was plagued by mistakes. The bomb missed its gaudy target by approximately eight hundred meters, destroying

In July 1946, the United States detonated the world's fifth atomic device—this time underwater—in a test called Operation CROSSROADS Baker. The twenty-one-kiloton bomb—identical to the one dropped on Nagasaki—created a radioactive tsunami that contaminated dozens of test ships and the US service members ordered to clean them. Library of Congress

monitoring equipment; the number of staff trained to check for fallout was insufficient; and their Geiger counters were unable to detect plutonium and malfunctioned in the humid air.

Such problems bode poorly for the next test, Operation CROSSROADS Baker, an underwater detonation. Undeterred by experts who warned the experiment would trigger a large tsunami, the military repositioned eighty-four of the ships that had survived the first test and submerged a plutonium bomb in their midst.

The official government film of the explosion explained what happened next. The blast displaced millions of tons of ocean and sent out a thirty-meter tall "boiling foaming wall of water" in every direction. Accompanying the wave was a "highly lethal spray" that was "intensely radioactive," its potency magnified because "neutrons bombarded sodium in sea water," creating radioactivity "equal to hundreds of tons of radium."[34] Beneath the surface, the blast scooped out a crater ten meters deep and shifted 1.5 million cubic meters of sand and rocks.[35]

Almost immediately, scientists, many not wearing protective equipment, sailed toward the hypocenter to retrieve the test data. In the coming days, tens of thousands of men were dispatched to the smoldering ships in vain attempts to decontaminate them. Such work had never before been attempted, and sailors were ordered to try washing them down with soap or sulfuric acid, blasting them with sand, coconut shells, and coffee grinds, and using flour and charcoal to trap the contaminants. Few were supplied with safety gear or radiation-monitoring badges, and even when areas were believed to be safe, the malfunctioning Geiger counters again provided false readings and failed to detect plutonium.[36]

The lagoon was severely contaminated, forcing support ships anchored far from the hypocenter to relocate as radiation drifted toward them. It accumulated in the algae and barnacles, settling on the support ships' hulls, while onboard filtration units converted seawater to drinking water, removing the salt but none of the radiation. Appalled by what was happening, ten days after the test, the commander of the operation, Admiral William Blandy, stated, "This is a form of poison warfare."[37]

The US Navy towed the ships to Kwajalein, Guam, Pearl Harbor, and the US West Coast, where it continued attempts to decontaminate them, exposing thousands more men to high levels of radiation; eventually, the military realized most of the vessels would never be safe enough to sail again and scuttled them.[38]

In the spring of 1948, three more tests were conducted at Enewetak, totaling 104 kilotons and adding more contamination to the Pacific. The US government did not try to keep the tests secret—the Able shot was broadcast live on radio throughout the world, and even Soviet officials were invited to witness it. The tests were reported internationally, except in one country: Japan. Just as US authorities were silencing references to the bombing of Hiroshima and Nagasaki, the Pacific tests were censored so as not to "unnecessarily alarm public."[39] In 1954, however, Japanese civilians once again became victims of both US radiation and a government cover-up.

## FROM KILOTONS TO MEGATONS

Led by Manhattan Project scientist Edward Teller, the United States had been considering the creation of a thermonuclear bomb—aka, a hydrogen bomb—since 1942. Based on the principle of a two-stage explosion, whereby fission in an atom bomb triggered fusion in a secondary device, the concept raised the potential power of blasts from the thousands of tons of explosives' kiloton range to the megaton range, equivalent to millions of tons.

In January 1951, the US government had begun testing atomic weapons at its Nevada Test Site, but the potential size of thermonuclear experiments so worried scientists that they decided to conduct them in the Pacific. In

May 1951, the government carried out its first trial to test the concept, GREENHOUSE George at Enewetak. The blast measured 225 kilotons—three times higher than the biggest test of the United States to date and ten times more powerful than Nagasaki's Little Boy.

The following year, on November 1, 1952, the United States conducted its first full thermonuclear test, IVY Mike. Essentially a building-sized bomb, the detonation measured 10.4 megatons, scooped 80 million tons of soil into the air, spread fallout as far as Guam, and wiped off the map the test island Elugelap.

By this time, the Cold War arms race was well under way. The USSR had tested its first atomic bomb in 1949, contaminating villagers near the site in Kazakhstan. Then, overtaking the United States, in 1953, the USSR detonated the world's first thermonuclear bomb, which was small enough to be delivered by aircraft. It exploded with the strength of four hundred kilotons.

In 1954, the United States constructed its own deliverable thermonuclear device and was ready to test it on Bikini Island. On March 1, despite weather reports warning winds could blow radiation toward inhabited islands, US scientists decided to proceed with the detonation without evacuating those in the potential fallout path. The bomb exploded at a yield far greater than its creators had expected—fifteen megatons, approximately one thousand times the strength of the Hiroshima bomb. "A freight train carrying the bomb's equivalent in TNT would span the United States from Maine to California," stated Jonathan M. Weisgall. The CASTLE Bravo test was the largest US explosion in history, creating a crater almost two kilometers wide and more than seventy meters deep.[40]

As predicted, the winds carried the fallout far, blanketing eighteen thousand square kilometers with contamination. According to the Defense Nuclear Agency's own admission, it was the worst single incident of fallout in US tests.

At the firing bunker, almost forty kilometers from the hypocenter, service members were trapped by radiation and had to be rescued by helicopter. The crews of approximately ten US ships also were forced to seek safety below decks as their vessels were enveloped by fallout. On neighboring islands, residents lacked the protection of concrete, lead. and steel hulls. On the island of Rongelap, four centimeters of radioactive coral powder fell; thinking it was snow, children played with it. The dust burned islanders' feet and made them vomit; their fingernails and hair fell out, as they experienced exposure of approximately 175 rads, compared to a recommended *annual* level of 0.5 rads. It took two days for the military to arrive on Rongelap and relocate residents. The Manhattan Project's successor, the Atomic Energy Commission (AEC), quickly issued a statement playing down the islanders' exposure and denying they had suffered any burns.[41]

For decades, the United States only admitted that four areas had been affected by fallout from its megaton-range tests: Rongelap, Bikini, Enewetak, and Utrik. But then a 1978 study by the Department of Energy revealed that another eleven atolls or islands had also been contaminated, including Ujeland, where the Enewetak people had been moved, and Rongerik, where Bikinians had been shunted.[42]

Not only were Marshallese exposed to fallout from the Bravo test, but also approximately twenty thousand crewmen aboard about 850 Japanese fishing boats. The best known of these was the *Daigo Fukuryu Maru* (Lucky Dragon #5).[43] On the morning of the test, its twenty-three crew members were fishing outside the exclusion zone when the bomb went off. Seeing the flash, one member recalled, "The sun rose in the west," and then they heard the roar of the explosion. Soon the men experienced similar fallout to the Rongelap islanders: Ash fell like sleet onto their decks. "It wasn't hot; it had no odor. I took a lick; it was gritty but had no taste," one remembered. Arriving back at their home port of Yaizu, Shizuoka Prefecture, two weeks later, the crew was sick with advanced radiation poisoning, including nausea, diarrhea, and hair loss.[44]

The exposure of the *Lucky Dragon* made headlines around the world, triggering a US campaign of disinformation and smears. Its ambassador to Tokyo called the crew's symptoms "mild" and blamed "fuzzy-minded leftists," "neutralists, pacifists, feminists, and professional anti-Americans" for playing up the incident. Even though the US occupation had been over for two years, the United States asked the Japanese government to censor announcements on the incident. Meanwhile, the chair of the Joint Committee on Atomic Energy speculated the ship had been on a spy mission and the Japanese Foreign Minister said he'd heard ash had been sent to the USSR—both of which were lies. US authorities refused to cooperate with Japanese medical teams in helping them to treat the victims. Instead, the United States ordered the dispatch of the ABCC—still embedded with ex-Unit 731 scientists—the head of which declared the crew members would recover within a month.[45]

On April 10, the Japanese government, instead of lodging a protest with the United States concerning the exposure of its citizens, declared its support for the ongoing US multimegaton tests in the Pacific. Six months after the Bravo explosion, the *Lucky Dragon*'s radio operator, Kuboyama Aikichi, died; before his death, he said he wanted to be the last victim of atomic and hydrogen bombs. The reaction of H-bomb creator Edward Teller? "It's unreasonable to make such a big deal over the death of a fisherman."[46] Later, the unborn child of one of the other crew members died so deformed the doctor told him he was "better off not seeing it." In the following years, ten more crew members were killed by illnesses linked to their exposure.[47]

The *Lucky Dragon* incident sent shockwaves through Japanese society. A total of 683 boats landed contaminated fish, and more than four hundred tons of tuna were dumped due to fears of contamination. Radiation was also detected in the rain over Japan, in both vegetables and the seas.

Under US occupation, censorship had kept Japanese people in the dark about the bombings of Hiroshima and Nagasaki, but now the tests of even more destructive devices gave the public the belated opportunity to vent their anger. The *Lucky Dragon* incident triggered a mass social movement against nuclear weapons; Tokyo housewives began a petition calling for the tests to be halted, garnering 32 million signatures, one-third of Japan's population.[48]

Faced with such public backlash, the US and Japanese governments realized they needed to do something to placate it. In November, the United States agreed to pay $2 million to the *Lucky Dragon* crew members, hospitals, and fishing association—much lower than the damage, which had been estimated at $7.2 million. The payment came with two conditions: First, it was "condolence money," with no admission of legal responsibility, and, second, it was final and no further payments would be made.

Behind the scenes, however, the United States and Japan hashed out another agreement: the expedited introduction of nuclear energy to Japan, a plan lauded by the *Washington Post* as the best way to "dispel the impression in Asia that the United States regards Orientals merely as nuclear cannon fodder!"[49]

In the coming years, the CIA worked closely with Japanese conservatives—including suspected war criminals—to sell the benefits of nuclear power to the Japanese public. The folly of this became only too apparent in 2011 following the meltdowns at the Fukushima Daiichi nuclear power plant.

For the United States, there was one final lesson to be learned from the *Lucky Dragon* incident: The open storage of US nuclear weapons in Japan was impossible. As a result, it decided to stockpile them on Okinawa, then under military control, and aboard US Navy ships at mainland Japanese ports, a secret from the public but enjoying the approval of Japanese officials.

## "MORE LIKE US THAN THE MICE"

Although the 1954 Bravo detonation became the focus of public attention, between 1946 and 1958, the United States conducted sixty-seven nuclear tests in the Marshall Islands, with a combined yield of 108 megatons. These amounted to an "average of more than 1.6 Hiroshima bombs per day for the twelve-year nuclear test program." In comparison, the atmospheric tests conducted at the Nevada Test Site added up to 1.05 megatons; the Marshall Islands bore the overwhelming burden of Cold War US nuclear tests.[50]

The explosions permanently disfigured the region, destroying three of the twenty-five islets in Bikini Atoll, and at Enewetak, tests rendered 57 percent of its original area uninhabitable due to obliteration or contamination. Radiation permeated the biosphere, contaminating the soil, the sea, fish, animals, and plants.[51]

Starting in 1972, Bikini residents returned to their home island after being assured it was now safe to inhabit; however, in the following years, serious contamination was discovered in wells, vegetation, and coconut crabs. Health checks revealed their bodies contained elevated levels of plutonium, cesium-137, and strontium-90. In 1978, residents had to leave their island again.

The people of Rongelap suffered a similar fate. In July 1957, following US promises that their island posed no health risks, residents returned home. In 1978, a survey revealed the island was contaminated. Thus, after more than two decades of eating and breathing high radiation, in 1982, they decided to leave. Despite appeals for assistance from the United States, no help was forthcoming, so the islanders asked the charity Greenpeace to help. In May 1985, the *Rainbow Warrior* evacuated the three hundred inhabitants to the island of Mejato. Two months later, as the boat was preparing to monitor French nuclear tests in the Pacific Ocean, French government operatives sank it with a bomb in Auckland Harbor, killing one crew member.

Residents of the Marshall Islands have experienced elevated rates of illnesses that many experts believe are linked to their exposure to radioactive contamination. Following the Bravo test, seventeen of nineteen Rongelap children developed thyroid abnormalities, and one died from leukemia. In 1999, the American Cancer Society's journal described the islands' cancer rates as "extreme." Lung, liver, and cervical cancers were between three and forty times higher than in the United States—and researchers believed the actual rates might be even more serious because some cases were not reported. Exposure to radiation was not the only cause—adding to the health crisis were such "lifestyle factors" as malnutrition and alcohol consumption. Marshall Islanders have also suffered the birth of "jellyfish babies," which one resident described as follows: "One was like the bark of a coconut tree. One was like a watery mass that was not humanlike. Another was again like a watery mass of grapes or something like that."[52] Between 1954 and 1958, an estimated one-third of the babies born by women contaminated with fallout died in the womb; between 1969 and 1973, the number was still one in five.[53]

US government scientists dispatched to the region repeatedly downplayed both the risks of radiation and their impact on the local population. But they did not investigate birth defects; blamed thyroid abnormalities on causes other than radiation; and, when calculating exposure pathways, failed to include local diets, which consisted of food gathered from neighboring islands, where radiation levels remained high.

Just like the *hibakusha* of Hiroshima and Nagasaki, many Marshall Islanders felt as though the US government was treating them like guinea pigs, monitoring their health without affording them any actual treatment, with the sole purpose of improving nuclear warfare. US authorities encouraged them to return to contaminated environments so the effects of radiation could be researched. Classified AEC documents reveal comments wherein scientists welcomed the opportunity to study the Rongelap residents: "Now, data of this type has never been available. While it is true that these people do not live, I would say, the way Westerners do, civilized people, it is nevertheless also true that they are more like us than the mice."[54]

With the secret Project 4.1, more than five hundred Marshallese were given experimental injections of radioactive isotopes and operations without their consent. Such disdain for human health extended to members of the US military.[55]

## RUNIT AND THE TWO-HUNDRED-THOUSAND-YEAR DOME

In the early 1970s, the islanders of Enewetak threatened to take the United States to court if their atoll was not cleaned up—so Washington was forced to act. At first, it wanted to dump the contaminated soil and detritus into the sea, but increased public awareness of the risks of such methods prevented it from doing so; the United States decided to load the waste into a 106-meter crater that had been formed in 1958, by the Operation HARD-TACK Cactus nuclear test on one of the atoll's islands, Runit.

From the start, the work was slipshod. To save money and reduce the risk of lawsuits, service members were used instead of trained civilian cleanup crews; with budgetary concerns in mind, little safety equipment was supplied. Plans to line the crater with concrete were also shelved, and the waste was dumped straight onto the porous coral ground.[56]

Between 1977 and 1980, approximately four thousand soldiers worked on Enewetak. Robert Celestial was one of them. In 1977, he was an army sergeant ordered to Enewetak to help clean up what he had been told was "postwar debris." His job was to pump water from the crater and gather

*Between 1977 and 1980, four thousand soldiers packed radioactive waste into a nuclear crater on Runit Island; the US government promised its concrete dome would last two hundred thousand years, but it is already leaking and many veterans believe they were poisoned by their work.* Department of Defense

topsoil from surrounding islands and dump it into the hole. The only safety equipment provided was a pair of rubber boots and, occasionally, a cloth mask to cover his mouth. "We were young, and we were ignorant," he told me when we met in Guam in 2019.[57]

In 1978, members of US Congress expressed worries about safety conditions at Enewetak, but the Secretary of Defense assured them that troops were equipped with hazmat suits and respirators. To keep up the pretense, when journalists visited the cleanup site, soldiers were given full-body safety equipment and posed for the cameras—the only time most of the men saw such gear during their time on the island.

Celestial ate fish, lobster, and coconut crabs. He also recalls being asked to urinate into bottles and submit them to the army clinic. When a colleague tried tracking down the results in later years, he was told they had been lost. "We were guinea pigs," Celestial concluded.

By 1980, soldiers had dumped some 110,000 tons of contaminated soil into the crater and capped it with 358 concrete slabs. The Department of Defense assured the government of the Marshall Islands that it would last for two hundred thousand years, but in 2013, Department of Energy researchers found it was already leaking. Regardless, from day one, the operation had been a PR stunt designed to placate the Marshallese. The overwhelming majority of contaminated materials still lay nearby beneath the sea, and even when, as scientists predict, the dome ruptures and splits, the addition of its contents to the already toxic lagoon will make relatively little difference.[58]

Also abandoned are the approximately four thousand veterans who took part in the cleanup. The military denies their work exposed them to radiation, and unlike those exposed during the actual tests of the 1940s and 1950s, they are not eligible for automatic VA coverage. One unofficial survey of troops who worked on the most contaminated islands suggested a cancer rate of 20 percent; meanwhile, Celestial estimates only six hundred of the original four thousand soldiers are alive today.[59]

## THE INDEPENDENCE OF THE MARSHALL ISLANDS

For more than three decades, the Marshall Islands were part of the Trust Territory of the Pacific Islands, at the whim of the United States. In 1979, they achieved independence and, in 1986, entered into a Compact of Free Association with the United States, according to which the United States agreed it was responsible to "address past, present, and future consequences of the nuclear testing claims." In 1988, the Nuclear Claims Tribunal was set up to determine compensation for the islanders. Initially, the United States put $150 million into the tribunal's fund, but since then the money has run out, and the United States owes approximately $2 billion in unpaid claims, including $244 million for the contamination at Runit.[60]

Today, the homepage of the US Embassy for the Marshall Islands perpetuates its seven decades of denials, deception, and spin. "Misconceptions about the history of nuclear testing in the Pacific and the present situation in the Marshall Islands hinder the full understanding of the importance of the testing," it states. Then it explains how the United States has paid out approximately $604 million in compensation and continues to finance ongoing medical checks.[61]

On paper, the amount, exceeding $600 million, might seem praiseworthy, but it pales when set against the US military's $494 billion nuclear weapons budget for 2019 to 2028. At an expenditure of $136 million per day, the payment for the Marshall Islands would be surpassed by Friday. The lives of the Marshallese continue to be disrupted by US nuclear weapons, with Kwajalein Atoll playing host to the Ronald Reagan Ballistic Missile Defense Test Site, a target for rockets fired from California. The Marshallese are not permitted to stay on Kwajalein; instead, they live on the adjacent island of Ebeye, one of the most cramped slums in the world.[62]

Throughout the Marshall Islands, overcrowding has become a dire issue, and the problem is worsened by Rongelap and Bikini remaining uninhabited due to concerns about contamination. In 2016, a Columbia University study found Bikini still contaminated, with levels higher than those considered safe by the United States and the Marshallese. As for the islands of Nam and Runit, the same study concluded that it is "unlikely either will ever be fit for habitation" due to Bravo test debris on the former and the dome dump site on the latter. The team, which confined its checks to the shoreline, speculated that the interiors of these islands might be more highly contaminated.[63]

Despite these latest findings, the US Embassy homepage, as of May 2019, claimed the following: "Today, the greatest source of radiation being absorbed by all persons living in the world, including the RMI, is from natural sources."[64]

In 1946, the US military chose the Marshall Islands to conduct its Cold War tests because they were relatively isolated and scientists hoped the radiation would be safely absorbed by the vast Pacific, but they failed to grasp how far radiation would spread through the ocean, air, and biota.

Between 1945 and 1992, the United States carried out 1,054 detonations, but it was one of only eight countries that conducted nuclear tests. The Soviet Union exploded an estimated 715 devices, France 210, and China 45. These tests dispersed radioactive isotopes throughout the world. A 1991 report estimated that these shots would cause 430,000 global cancer deaths by 2000. In 1999, a study by the International Council for Science estimated that the total number of cancer deaths triggered by these tests might reach 2 million.[65]

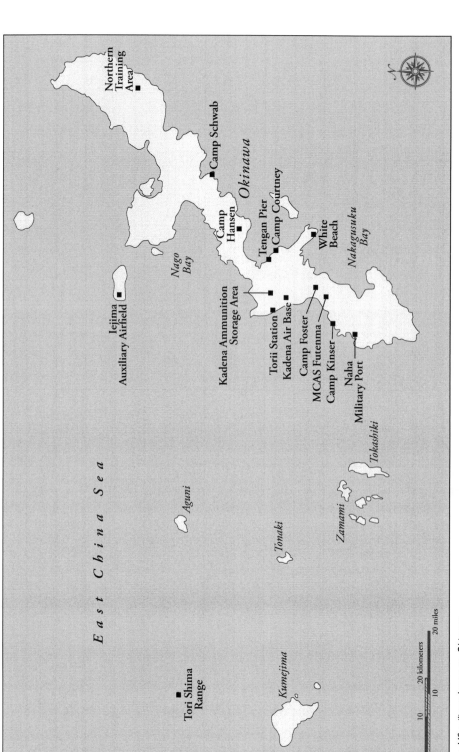

*Main US military bases on Okinawa*

# 3

# OKINAWA

## "The Junk Heap of the Pacific"

The fifty-five islands of the Ryukyus stretch from the southern tip of Japan to the coast of Taiwan. Throughout the middle ages, they were famous in the region as a trading kingdom, with Ryukyu ships plying the ports of Southeast Asia to collect ivory, spices, and woods in exchange for ceramics from China and lacquerware from Japan. Beginning in the 1300s, the Ryukyus had a tributary relationship with China, according to which the two nations shared their expertise, goods, scholars, and skills, including what would later become karate. It was a profitable relationship for both China and the Ryukyu Kingdom—and the islands' politics and culture were heavily influenced by the mainland; however, such prosperity brought it the unwanted attention of Japanese pirates and samurai; in 1609, the Satsuma clan invaded the Ryukyu islands and demanded a slice of the moneys brought in from the tributary trade with China. For the next two centuries, the Ryukyu Kingdom was compelled to pay off both China and Japan, causing a heavy economic burden on the island. With Japan closed off from the world, samurai spied on the Chinese envoys to the Ryukyus and concealed themselves among Ryukyuan visits to China—but to avoid disrupting the lucrative relations, Japan deferred from formally annexing the islands.[1]

This all changed in 1853, when US gunships anchored off Okinawa on the way to Tokyo to force the closed nation to open to trade. The arrival rattled Japanese rulers, who realized they were on the brink of colonization, like the rest of Asia, so they embarked on a rapid program of reforms accompanied by the consolidation of Japan's borders. In 1869, Japan annexed Hokkaido to its north and, ten years later, abolished the Ryukyu Kingdom, turning it into a prefecture, Okinawa. Mainland leaders saw themselves as superior members of the Yamato bloodline and harbored doubts as to whether Okinawans could be molded into civilized subjects. Japanese officials began to impose mainland culture, banning local languages and such customs as female tattooing and shamanism. Japan also imposed extortionate taxes, contributing to a famine that devastated the islands in the 1920s. Poverty forced some Okinawans to emigrate to the mainland, where, along with Koreans, whose nation had been occupied since 1910, they experienced discrimination from landlords and employers; others headed to toil on sugarcane plantations on the Micronesia islands of Saipan and Tinian, which Japan had seized from Germany following World War I.

By 1944, Japanese leaders realized they were losing World War II; the only questions left were what shape their surrender would take and how they could preserve the emperor as the nation's head. If they could delay an Allied invasion of Tokyo by stalling US troops on Okinawa, it would buy time to reinforce mainland defenses and, if the fighting on Okinawa were bloody enough, maybe even turn US public opinion against the war. In the vocabulary of the game of *go*, the Japanese military saw Okinawa as a *sute ishi*, a stone to be sacrificed for the greater good. On Okinawa, the Japanese military forced civilians to construct runways and defenses, and then, with US forces drawing nearer, conscripted women and children into nursing and combat platoons.

## THE STORM OF STEEL

On March 26, 1945, the United States attacked Okinawa's outer Kerama islands, followed by its main invasion on April 1. By the end of fighting in late June, more than 70,000 Japanese and 14,000 American troops had been killed; 25,000 GIs were incapacitated by shellshock, a testament to the ferocity of the combat. More than a quarter of Okinawa's population—140,000 people—were killed. Included among the dead were at least a thousand civilians executed by Japanese troops as "spies" for attempting to surrender or speaking the banned Okinawan languages, and many families were forced by Japanese troops to commit suicide to avoid the shame of capture.[2]

For Okinawans, the battle was both genocide and ecocide. Combat churned farms to wastelands; leveled forests; and clogged streams and springs, the few sources of fresh water. During the battle, the US military expended two hundred thousand tons of heavy munitions, approximately 5 percent of which failed to detonate. Since 1945, 710 people have been killed and 1,281 injured by this unexploded ordnance (UXO). An estimated two thousand tons of UXO remain beneath Okinawa's ground, and experts estimate it will take seventy more years to clear.[3]

Often overshadowed by the explosive dangers of UXO is its chemical contamination. World War II munitions contain trinitrotoluene (TNT) and cyclonite, both of which are toxins. TNT damages the liver, blood, and immune system, and causes reproductive problems; cyclonite harms the neurological system and causes seizures. For decades, this ordnance has corroded, seeping its contents into Okinawa's environment. The toxic effects of ordnance were apparent even as combat was waged; US troops reported a "greenish smoke" emanating from Japanese shells when they exploded, causing nausea and vomiting. The smoke was attributed to picric acid, one of the toxic ingredients of explosives.[4]

During the Battle of Okinawa, both the United States and Japan employed a variety of unconventional ordnance. On the American side, this included napalm and chemical smoke grenades. A mixture of gasoline and jelly, napalm sticks to skin and burns even when submerged in water; people caught near a napalm explosion can die from suffocation due to its combustion of available oxygen. During the Battle of Okinawa, US aircraft dropped hundreds of thousands of liters of napalm from the air, while tanks sprayed it from their barrels. To force enemy soldiers and civilians from their underground hiding places, the US military also used toxic smoke grenades. Primarily designed for use outdoors to hide troop maneuvers, the use of such grenades in confined spaces can be fatal. Accounts from Okinawan civilians trapped in caves recalled their horror as canister after canister of smoke tumbled into their hiding places, suffocating many to death.[5]

The Japanese military also deployed at least two types of chemical weapons on Okinawa, but probably due to President Franklin D. Roosevelt's threats of "retaliation in kind," there is no evidence of their use against Allied troops. Brown Agent, hydrogen cyanide, kills by attacking the ability of the brain, heart, and lungs to use oxygen. For military use, workers at Okunoshima injected it into small glass grenades and added copper powder as a stabilizer. In July 1998, one of the grenades was discovered in the ruins of an army shelter in Okinawa's Itoman City; the glass ball was ten centimeters in diameter and still contained copper powder. Sealed in glass, hydrogen cyanide can remain deadly for decades and poison anyone unlucky enough to crack open its fragile casing.[6] The other chemical weapon was Red Agent. An after-action report noted the discovery of a "small quantity" of vomiting gas cannisters, but there were no accounts of their use against US troops.[7]

In Japan, the Battle of Okinawa is often referred to as a storm of steel, but a more accurate description might be a chemical storm of TNT, toxic smoke, and hydrogen cyanide. Sadly, for the civilian survivors of the war, the poisoning of their island had only just begun.

## THE JUNK HEAP OF THE PACIFIC

Whereas the Japanese military envisaged Okinawa as a *sute ishi*, the United States saw the island as a stepping-stone toward the invasion of mainland Japan, which was projected to involve millions of Allied troops. In preparation for this assault, the United States herded Okinawan civilians into internment camps so it could turn their land and former Japanese bases into new US military installations. On August 9, for example, the B-29 Bockscar made an emergency landing at an airfield on Yomitan after running out of fuel following its atomic bombing of Nagasaki.

After Japan's surrender, the military's attention shifted to Tokyo and its mainland occupation, neglecting Okinawa and the civilian population there. In the internment camps, at least sixty-four hundred people died from malnutrition and malaria; survivors, having lost everything, depended on US trash and surplus to make their clothes and cooking utensils, as well as the shelters in which they lived. Beginning in 1946, the United States began to permit civilians to leave the camps, but upon arrival at their home communities, many Okinawans found the military had built a base on their land. Meanwhile, US troops ran amuck; during a six-month period in 1949, they committed 29 murders, 18 rapes, 16 robberies, and 33 assaults against the island's 600,000 residents.[8]

A US Army documentary from the 1950s—"Okinawa: Keystone of the Pacific"—provides a rare environmental snapshot of the island at this time: "Okinawa after the war was a forlorn place. Equipment shot up during the war and equipment piled up for the invasion of Japan lay around to rot and rust away. The rock was forgotten—now it was called the junk heap of the Pacific."[9]

From this period comes the first account of victims of US military contamination on Okinawa. During and after the war, US troops had rounded up the residents of Iheya Island and relocated them to an internment camp; with the civilian population out of the way, the military had built an ammunition storage area. Following the closure of the internment camps, the residents returned to their island. One of them was seventeen-year-old Nishie Yuki and her family of eight.[10]

Arriving at their village, they found their home destroyed, so they moved to nearby land and began drawing water from its well. In about 1947, Nishie's family fell ill. First, their eyes started to sting, and their skin turned dark brown, mottled with spots. They developed numbness in their hands and feet. As the symptoms spread, they became bedridden, and their stomachs swelled with fluid.

First to die was Nishie's father. Next was her older sister. Within a year, all eight members of her family had died from the same symptoms. Worried the illness was infectious, villagers shunned their funerals. Nishie expended the family savings on hospital visits to mainland Okinawa, but doctors were unable to diagnose the cause of her illness. Away from home, she'd slowly recover her health, but upon returning to Iheya she'd fall sick again.

During one of her prolonged visits to a hospital, another family had rented the property, and they too fell ill. Suspecting the root of the illnesses might be environmental, the authorities drew water from the well near her home. Tests revealed it was tainted with high levels of arsenic, but it was not until 1971, that the cause of the contamination was discovered, when two empty metal canisters dated to the US military's occupation of the village were excavated near the well.

## BAYONETS AND BULLDOZERS

In 1941, the United States had signed the Atlantic Charter, in which it pledged not to seek any territorial gains from World War II. After Japan's surrender, however, the US military saw its Pacific conquests in Okinawa and Micronesia as war booty won at the cost of millions of dollars and tens of thousands of American lives. Washington conceived a militarized Okinawa as a means to counter the spread of Communism in Asia and within Japan itself. In 1947, Emperor Hirohito gave his seal of approval to US plans for Okinawa, telling the Americans they could keep the island for ninety-nine years.[11]

In 1952, the Treaty of San Francisco ended the US occupation of mainland Japan, but it cemented US jurisdiction of Okinawa, granting the military control of the "territory and inhabitants of these islands, including their territorial waters." Even now, Okinawans remember April 28, 1952, as a "Day of Shame." Japan had sacrificed their island in 1945, to protect the imperial system, and now, seven years later, they had been betrayed again. Between 1952 and 1972, Okinawa existed in a geopolitical gray zone, a US military colony protected by neither the constitution of the United States nor that of Japan.

In the following few years, a period known by Okinawans as the time of bayonets and bulldozers, the US military forcibly seized more land to expand its current bases and create new ones. As if to remind civilians of their defeat, the military named many of these facilities after American Battle of Okinawa heroes—for example, Camp Foster, Camp Hansen, and Camp Schwab. According to US government records, by 1955, military land seizures had displaced 250,000 residents out of a total population on the main island of 675,000.[12]

To construct Naha Air Base, for example, the military drove residents off their property using tear gas, the first of many Okinawan victims of the substance in the years to come. On Iejima Island, the military seized land to construct an aircraft bombing range. With their farms destroyed and no way to support themselves, villagers resorted to gathering waste metal from the base. Between 1959 and 1961, at least eleven islanders were killed or wounded while collecting or dismantling American ordnance; one died when a dummy hydrogen bomb exploded during military training operations.[13]

Elsewhere, Okinawans picked through the detritus of war, scavenging chunks of broken Japanese and US military equipment to earn a living in what became known as the scrap boom. In 1956, the sale of this metal outstripped sugar as Okinawa's largest export. Much of this scrap was exported to Japanese factories as the mainland's economy, unlike Okinawa's, began to recover—spurred by the Korean War, during which US military contracts pumped billions of dollars into Japanese firms. Poverty forced many Okinawans to make a stark choice: emigrate to Japan or overseas, or work on the bases. Thousands fled to South America, where the arable land they'd been promised turned out to be jungle. Work on the bases was little better; Okinawans were assigned the most dangerous jobs while suffering according to a discriminatory wage scale that pegged their salaries below those of Japanese and Filipino laborers.[14]

Okinawans were powerless to protest these injustices. Unlike on mainland Japan, where US occupiers had introduced a new constitution encouraging a veneer of democracy, Okinawa was a dictatorship under the control of a US military high commissioner. In the early years of US occupation, Okinawans had no say in the governance of their island, the press was tightly controlled, and unions were banned. Starting in 1957, the CIA, well established on Okinawa by this time, channeled money into the Liberal Democratic Party—the start of decades of sabotage it conducted to stifle the free growth of democracy on the island. The CIA was headquartered at Camp Chinen and flew Air America missions from Hamby Air Strip. Among its Okinawa operations was an attempt to train anti-Communist Chinese guerillas to overthrow Mao's government; the rebels tricked the agency out of $50 million, and the uprising failed to materialize.[15]

By the late 1950s, the US government had built approximately 140 military installations on Okinawa, occupying almost 30 percent of its land. Many of these bases were for the United States Marine Corps (USMC), which had been relocated from mainland Japan due to local opposition to its presence there. The Pentagon prioritized its own infrastructure at the expense of such civilian projects as schools and hospitals; fuel pipelines crisscrossed the island, cutting through communities. This use so blurred the distinction between military and civilian areas that, in 1959, one US report surmised, "The southern half of Okinawa does not really contain a series of bases but should be classified as a single military base complex."[16]

In addition to smothering residential districts, agricultural land, and cemeteries, the bases also occupied areas vital to Okinawa's ecosystem, particularly sources of fresh water. In 1957, the military set up the Northern Training Area (NTA) in the Yanbaru jungles, a region described as Okinawa's water jug because it provides the island with a reliable source of drinking water. Farther south, Kadena Air Base, the largest United States Air Force (USAF) base in Asia, sat upon dozens of fresh water springs and wells. Other installations, including Makiminato Service Area, Camp Courtney, and Camp Schwab, occupied beaches that had long been used as a source of fish, shellfish, and seaweed by Okinawans.

## THE VIETNAM WAR ON OKINAWA

Throughout the early 1960s, the United States had been deploying thousands of military advisors to bolster its puppet regimes in South Vietnam. On March 8, 1965, it sent its first combat troops; members of the Third Marine Division came from Okinawa. Less than forty-eight hours later, Okinawans suffered the first chemical impact of the new war. On March 10, as pupils of Ginoza Middle School sat down to lunch, they felt a sharp pain in their throats, followed by coughing and difficulty breathing. Teachers ordered the evacuation of the school, and once outside, they discovered a group of marines on nearby military land. The teachers demanded to know what was happening to their students, but the Americans dismissed their concerns and told them their pupils must have accidentally inhaled gun smoke. It took three days of pressure from the school, local authorities, and the media to find out what had really occurred: Soldiers, training on the USMC base, had allowed CS gas to drift into the school.[17]

In the following months, as the United States became increasingly involved in the war in Vietnam and then Cambodia and Laos, Okinawa experienced a surge of troops and supplies not seen since the planned invasion of mainland Japan two decades prior. For Washington, Okinawa was in the perfect position to wage war in Southeast Asia—far enough from the front lines to be protected from enemy attack but close enough to serve as a staging post for troops and supplies. Most importantly, the island was under military control, where the Pentagon could operate freely without civilian encroachment.

According to military estimates, 75 percent of the supplies its troops used in Vietnam passed through the island, mostly via White Beach and Naha Port, where the US Army's Second Logistical Command and Okinawan stevedores worked twenty-four hours a day, offloading ships from the United States. Other material arrived at Kadena Air Base, which, between 1965 and 1973, handled one million flights, approximately one every three minutes. At the peak, there were fifty thousand Okinawans employed on military bases, including in the NTA, where they were paid a dollar a day to dress up as Southeast Asian peasants and add a touch of realism to simulated Vietnamese villages.[18] So great was Okinawa's importance to the US military posture that, in December 1965, Admiral Ulysses S. Grant Sharp, commander of US Pacific Forces, stated, "Without Okinawa, we couldn't continue fighting the Vietnam war."[19]

Between 1965 and 1975, seventeen Okinawans were murdered by Americans, and others were raped or robbed; service members mowed down civilians in hit-and-runs. Perpetrators were shipped back to the United States for trial and, in some cases, released without charge. Okinawans also bore the brunt of aircraft crashes.[20]

The island's environment was severely damaged by the influx of troops and intense use of the bases. Waste oil, detergents, and sewage spilled into rivers and shorelines, ruining the fishing industry, and fuel leaked from the pipelines routed through civilian communities.[21] By June 1968, Kadena Air Base had so polluted local wells that water drawn from them could be lit with a match, and one source near Marine Corps Air Station (MCAS) Futenma earned the moniker "Gasoline Spring" due to its smell.[22] The military cared little about the impact on civilians.

When one Okinawan leader complained about fuel spills in 1974, the US Consulate in Naha sent a classified cable to the State Department explaining that the "pipeline has now been added to leftist catalogue of evils of US base system."[23]

Interviews with veterans reveal the consequences of running industrial-scale operations under lax safety standards with no care for efficiency or the need to reduce waste. The US military's fleet of thousands of jeeps, buses, and trucks dumped tons of radiator coolant, used oil, and battery acid at maintenance sites, including the US Army's Makiminato Service Area. Also, at Makiminato, the Seventh Psychological Operations Group poured down drains photograph developing fluid and the inks from the presses it used to print Vietnamese propaganda leaflets. The military's main mortuary at Kadena Air Base, infamous for routing heroin from Thailand to the United States, flushed excess formaldehyde down sewers. Meanwhile on-base laundries went through tens of thousands of liters of dry-cleaning chemicals and detergents. Added to this toxic cocktail were the substances Americans demanded to live the semblance of a comfortable First World existence while on subtropical, poverty-stricken Okinawa: the dichlorodiphenyltrichloroethane (DDT) sprayed weekly from trucks to kill mosquitoes, the malathion to keep military and civilian dogs free from fleas, and the pesticides and herbicides used to maintain on-base lawns and golf fairways.

On Okinawa, the military disposed of its waste at sea, or by burning, burial, or sale to unsuspecting locals. Photographs from October 1964 record gas-masked US troops disposing of unspecified "chemical material" from the back of an army truck aboard a boat into Okinawa's sea. Reports from the Second Logistical Command also catalog the dumping of tons of damaged ammunition at sea and the burning of hundreds of kilograms of ordnance—a dangerous practice for communities located downwind.[24]

Burial was another common disposal method with such waste as pesticides, herbicides, and tetraethyl lead, an additive in fuel that wrecks the human central nervous system and can cause reproductive and developmental disorders. In the early 1970s, Okinawan base worker Tamura Susumu witnessed the burial of a surplus stock of what his American boss described as "coal tar." Thirty years later, after the land had been returned to civilian use, the 187 barrels were unearthed by a construction crew. The barrels' contents, a sticky black liquid, had leaked from their containers, and the discovery sparked a public outcry. Local authorities conducted only perfunctory tests before declaring them nontoxic and sending them away for disposal. At first, the US military denied the barrels were theirs, but

*Photographs from October 1964, show men in gas masks dumping unspecified "chemical material" into Okinawa's sea from the back of a US Army truck aboard a US Army boat.* US National Archives

thanks in part to Tamura's testimony, they finally admitted responsibility. Nevertheless, due to the long-standing US–Japan agreements explained in chapter 5, the cleanup bill of 20 million yen ($182,000) was footed by Japanese taxpayers.[25]

During the Vietnam War, the final way in which the US military disposed of surplus chemicals was via auctions and sales. According to one resident, the military sold surplus stocks of Vietnam War herbicides to a local municipality, which then sprayed them around the community to clear vegetation. Veterans also recalled the sale of such chemicals to local farmers who valued their powerful weed-killing properties. In 1971, a private company bought a large stock of herbicides from the military and subsequently dumped them on land in the Haebaru and Gushikami districts. The chemicals, which contained poisonous pentachlorophenol, leaked into a nearby river, resulting in contamination of local tap water. Local schoolchildren fell sick with stomach pains, and the water supplies of thirty thousand people were disrupted.[26]

Chemical leaks on the bases injured many Okinawans. In August 1975, a large spill of industrial detergents exposed base workers to lead, cadmium, and hexavalent chromium, a substance that can cause lung cancer, at levels eight thousand times the safe limit. The US Consulate in Naha wrote a

Okinawa officials visit a dumpsite of surplus US herbicides in southern Okinawa in 1971; the substances leaked into a nearby river and sickened children. Okinawa Prefectural Archives

secret cable to the State Department dismissing the accident as a "flap" and writing, "The newspapers and the leftists will certainly make good use of this issue against us."[27] Another secret cable from September 1975 stated that the issue of base pollution was "readily exploitable by the leftists against American interests on Okinawa" and the US government ought to prepare the "best possible political posture in the event of future pollution accidents."[28]

This comment highlights the disdain US authorities had for those poisoned—an approach consistent with the one taken in the aftermath of Hiroshima and Nagasaki, and honed throughout the Bikini nuclear tests. The military did all it could to hide the truth about contamination, and whenever news escaped, it attempted to downplay the dangers or discredit those who revealed it; however, unbeknownst to Okinawans and most US service members, there was an even larger risk present on the island.

## US WEAPONS OF MASS DESTRUCTION ON OKINAWA

During the twenty-seven-year US occupation, Okinawa possessed one of the most concentrated stockpiles of weapons of mass destruction on the planet: approximately one thousand nuclear warheads and thirteen thousand tons of chemical weapons. In the 1960s, the military conducted operations on the island under its human test program, Project 112, as well as experiments with biological weapons on at least three sites. At the time, incidents involving these weapons sickened Americans and local civilians; today, Cold War burials and sea dumps of these munitions continue to threaten health on the island.[29]

## US NUCLEAR WEAPONS ON OKINAWA

On mainland Japan, the public backlash against nuclear tests following the 1954 contamination of the *Lucky Dragon* convinced the United States that stationing such nuclear weapons in Japan was politically unfeasible—but the situation on Okinawa was different. "Here there are no restrictions imposed by a foreign government on our rights to store or to employ atomic weapons," stated the Melvin Price Report, penned by a special subcommittee of the House of Representatives Armed Services Committee in 1955. Despite this and other tacit admissions, the US government maintained a policy of neither confirm nor deny, which kept the deployment of nuclear weapons to Okinawa secret for decades. Only in recent years has the declassification of top-secret documents, combined with testimonies from US veterans, made it possible to grasp the scale of nuclear weapon storage on Okinawa and some of the accidents involving them.[30]

According to documents released by the Pentagon to US researchers via the Freedom of Information Act (FOIA) in the late 1990s, the first nuclear weapons were brought to Okinawa in 1954. In the following years, approximately twelve hundred warheads were kept at various sites, including Naha Air Base, Kadena Air Base, and army ammunition depots at Chibana and Henoko. There were at least eighteen types of nuclear weapons on Okinawa, such as megaton-range bombs, rockets, and smaller tactical shells designed for use on the battlefield.[31]

In 1955, the US Army brought six atomic cannons to Okinawa that could fire small 280mm nuclear munitions, each packing about the same power as the Hiroshima bomb, thirty kilometers. During its first test firing of a nonnuclear shell in Ginoza Village in October 1955, the blast shattered fifty windows in a school 150 meters away, injuring four students by flying glass. At the same time, the United States also deployed Honest John rockets, capable of delivering nuclear warheads or 356 cluster bombs, each filled with a half-kilogram of nerve agent.[32]

In the 1950s, the USAF deployed F-100 fighter-bombers to Kadena Air Base with 1.1-megaton thermonuclear bombs. One of those to visit Kadena was Daniel Ellsberg, at the time an adviser for the RAND Corporation, analyzing the military's nuclear weapon deployments. In his 2017 book *The Doomsday Machine*, he described the slipshod operations he saw on Okinawa. The 1.1-megaton bombs were too large to fit inside the F-100 aircraft, so they had to be slung beneath their bellies. The bombs themselves lacked the full safety features of other devices, meaning if they were accidentally dropped or involved in an accident, there was the risk of a partial or full detonation.

Okinawans detested the F-100 for another reason: In June 1959, one of the jets had crashed into Miyamori Elementary School in Ishikawa City (present-day Uruma City), killing twelve pupils and six other civilians. The pilot parachuted to safety and subsequently enjoyed a long and distinguished career in the USAF.[33]

Additionally, the military built four silo complexes on Okinawa holding a total of thirty-two MACE missiles, which, armed with 1.1-megaton warheads, had a range of two thousand kilometers enabling them to strike China and the Soviet Far East.

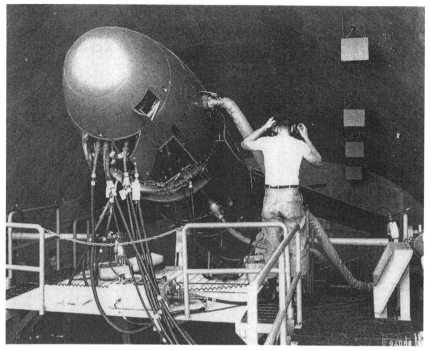

*During US occupation, Okinawan bases possessed an arsenal of more than one thousand nuclear weapons; here US Army technicians conduct maintenance on a Mace missile—capable of carrying a one-megaton warhead—in 1962. US National Archives*

In 2011, I conducted the first interviews with USAF technicians in charge of these missiles, and they described the secrecy surrounding their deployment to Okinawa. The Secretary of Defense had banned them from using the word *missile* on their uniform patches, and the weapons themselves were always transported at night, hidden beneath tarpaulins.

In 1962, just after the Cuban Missile Crisis, a potentially catastrophic incident occurred in the bunker in Onna, when one of the missiles developed an electrical malfunction. With warning alarms blaring throughout the underground complex, one technician climbed into the firing tube to rewire the missile. It took six hours before he finally managed to resolve the problem. "Had it detonated, the explosion would have set off the seven other missiles alongside. The blast itself would have been catastrophic, and the fallout would have left the island uninhabitable for decades." All the technicians I interviewed agreed that the presence of nuclear weapons on Okinawa made it a key target in a preemptive or a retaliatory strike; one of the veterans described Okinawans as "human shields."[34]

Even though Okinawa and the rest of the world was spared annihilation during this period, there were at least two incidents—"Broken Arrows"—that dumped nuclear weapons into the seas near Okinawa. In the 1950s, the US Army deployed its first Nike missile systems to eight sites on the island. The thirteen-meter-long weapons were equipped with warheads designed to knock incoming enemy aircraft out of the skies. In June 1959, US veterans recalled, an accident took place at Naha Air Base. With a warhead attached and the missile in a horizontal position, technicians were conducting electrical tests on the system. A short circuit occurred, igniting the engines and firing the missile into the sea; one soldier was cut in half by exhaust blast, and another later died from his burns. According to one of the surviving soldiers, the weapon was later recovered in secret by the military. The United States still declines to comment on the incident.[35]

The second Broken Arrow sunk into Okinawa's waters in December 1965. The aircraft carrier USS *Ticonderoga* was sailing from Southeast Asia to Yokosuka Naval Base when one of its aircraft rolled overboard, carrying its pilot and a B-43 one-megaton bomb to the bottom of the sea. The Pentagon kept the incident secret for the next sixteen years. In 1981, upon finally revealing what had happened, the US government claimed the incident had occurred more than 800 kilometers from land. Navy documents later revealed the accident had actually taken place about 130 kilometers east of the Ryukyu islands. Despite the incident providing incontestable proof the US Navy was bringing nuclear weapons to Japan, Tokyo did not lodge any form of protest. The bomb remains on the seabed, slowly corroding, potentially leaking radiation, and susceptible to disturbances by unsuspecting trawlers and tectonic shifts.[36]

As well as the dangers of the actual weapons, during the US occupation there were also other incidents involving radioactive contamination on Okinawa. In the Vietnam War, US nuclear-powered submarines frequently visited Naha Port, White Beach. and US Navy bases on mainland Japan. In August 1968, one of these vessels leaked radioactive cobalt-60 into the waters of Naha Port, and three divers claimed to have been sickened by their exposure to the substance, which had accumulated in mud at the bottom of the harbor. In May 1972, cobalt-60 was again detected at Naha Port, and this time in shellfish at White Beach, too.[37]

Records from the Second Logistical Command, the group tasked with handling supplies for the military on Okinawa, also described the disposal of radioactive waste. The material, stated one 1970 report, was "generated in or evacuated to Okinawa" and related to "ammunition operations." In the United States at this time, it was standard operating procedure for the military to bury such waste on its bases.[38]

A CIA memo from July 1969, dismisses residents' fears of radiation as irrational:

> With their well-known "nuclear allergy" as a basis, the Japanese are easily aroused by scare reports of radioactive pollution of their waters, and of other "evils" associated with US bases in Japan and Okinawa. . . . These fears and irritants are always present among the Japanese, and are readily susceptible to exploitation by anti-US elements on the slightest pretext.[39]

Throughout the 1960s, Okinawans demonstrated for the reversion of their island to Japan, where they hoped the constitution would guarantee them democratic rights and a reduction of the military presence, particularly the removal of the nuclear weapons they suspected were housed there; mainland Japanese, the majority of whom opposed the war in Vietnam, supported these demands for reversion. In 1969, former US ambassador to Japan Edwin Reischauer broke a long-standing taboo and referred to Okinawa as a "colony of one million Japanese." By the late 1960s, the reversion movement had become so vociferous that the United States feared its military bases on Okinawa might be at risk, so it agreed to return Okinawa to guarantee the security of its bases, a deal that left the US presence intact and allowed continued use of the bases without constraints. For Tokyo, the agreement provided Japan with the ongoing security of the US military without having to increase the US presence on the mainland. Once again, the Japanese government displayed its obsequiousness when it agreed to pay the United States approximately $685 million—including at least $70 million for the removal of nuclear weapons—in a deal Tokyo hid from the public for decades. In addition to this one-off payment, the United States received long-term financial benefits because following reversion, the Japanese government would pay for the upkeep of the bases, including the salaries of base workers and the facilities' electric and water costs.[40]

Outwardly, the Japanese government pledged to Okinawans that reversion would take place according to the principles they had long demanded—*kaku nuki hondo nami* (no nuclear weapons and a base presence proportionate to the mainland)—but both promises were broken. In the following years, the burden of bases on the island would increase compared to the mainland; in 1972, 42 percent of the land used by the US military was on the mainland vis-à-vis 58 percent on Okinawa, but today the island hosts roughly 70 percent of the US military bases in Japan. Moreover, when Okinawa reverted to Japanese control on May 15, 1972, there were still eight different types of nuclear weapons on Okinawa. Department of Defense records show these were finally removed the following month; the United States has never commented on their removal.[41]

The US military's use of Okinawa as a nuclear weapon storage site did not end in 1972—during reversion negotiations, the Japanese government agreed to allow the US military to return nuclear weapons to Okinawa in an emergency.[42] The deal, which remained hidden for decades, blatantly violated Japan's nonnuclear principles. It again shows the Japanese government's willingness following reversion to treat the island as territory where actions could be taken that would not be allowed elsewhere in the nation; since the 1972 "reversion," Okinawa has suffered from the double subjugation of both Washington and Tokyo.

## BIOLOGICAL WEAPONS TESTS AND PROJECT 112

During the same period that the Pentagon was packing Okinawa with nuclear warheads, it was also conducting biological and chemical weapons tests there. According to reports from Fort Detrick, between May 1961 and September 1962, military scientists conducted eleven biological weapons experiments in Shuri (Naha City), Ishikawa, and Nago City. The experiments involved rice blast, a fungus that can rot entire fields of rice and, if it reaches epidemic proportions, the harvests of whole regions. On Okinawa, scientists used a midget duster to release spores alongside fields; three varieties of rice were infected.[43]

The experiments were part of US research into the disruption of food supplies in East Asian countries and designed for potential deployment in future wars there. Biological weapons expert Sheldon H. Harris asserts the research on Okinawa was so successful it encouraged the US military to embark on tests with a thousand more substances.[44]

According to one US Army report, in 1962, the United States also began to conduct operations on Okinawa under the auspices of Project 112. Described by the Department of Defense as "biological and chemical warfare vulnerability tests" around the world, Project 112 exposed thousands of unwitting American service members to such substances as sarin and VX agents. The program ran from 1962 to 1974, and for decades after, the Pentagon denied its existence; however, in response to illnesses among its human test subjects, it was forced to admit to the program in 2000. Two years later, Congress ordered the Pentagon to create a register of service members exposed during Project 112, and it released a list of the project's locations—but Okinawa does not appear on it.[45]

US Army reports show Okinawa was a site for Project 112 as early as December 1962. The site, known as Red Hat, was managed by the US Army 267th Chemical Company, stationed at Chibana Army Ammunition Depot. Before coming to Okinawa, the report stated, the thirty-six-member platoon had received training at Rocky Mountain Arsenal, Denver, the Pentagon's chemical and biological weapons facility. Another report released by the National Archives and Records Administration via a FOIA request also confirms the existence of the operation on Okinawa, calling it "DOD Project 112 (RED HAT)."[46]

The primary task of the army company was the management of a chemical weapons stockpile, but there is evidence it also conducted tests on Okinawa. Included in the Department of Defense's list of those exposed during Project 112 tests is Don Heathcote, a former marine stationed at Camp Hansen in 1962. Heathcote's Project 112 file states, "Sprayed from numbered containers." Heathcote clearly recalls the circumstances of the tests. Assigned for approximately one month to the NTA, he was ordered to conduct trials with chemicals from color-coded drums. Without safety equipment and closely monitored by a senior officer, he repeatedly sprayed foliage with different substances. The chemicals killed the jungle, says Heathcote, and also damaged his health. "Soon after I returned home, I underwent an operation to extract polyps from my nose. The doctors removed enough to fill a cup. Plus, they diagnosed me with bronchitis and sinusitis connected to chemical exposure."

Other marines recall similar spray tests on Okinawa during this period that left them with such chronic breathing problems as scarring of the lungs and neurological illnesses, including Parkinson's disease. One of them, Gerald Mohler, fell seriously ill soon after being ordered to camp in a defoliated forest in 1961. "Were we marines used as guinea pigs on Okinawa? I think so," said Mohler in an interview in April 2012. He passed away the following year, but the US government still has not admitted its responsibility in exposing him and his colleagues to chemical experiments on the island, nor does the US government include Okinawa on its official list of places where Project 112 operations were held, frustrating veterans' claims for VA support.[47]

## CHEMICAL WEAPONS ON OKINAWA

The deployment of chemical weapons to the Red Hat site was conducted under the same intense secrecy as the storage of nuclear weapons; however, thanks to a combination of declassified documents and veterans' testimonies, it is possible to reconstruct the details of the US chemical weapons program on Okinawa.

The US Army first brought chemical weapons to the island in 1953, when it deployed a small stockpile of mustard agent. The following year, it brought sarin agent. First invented in 1938, sarin was one of the new generation of nerve agents, far deadlier than the blister and choking chemicals used in World War I. Initial exposure to sarin causes pupils to pinpoint, leading to visual problems, followed by drooling and sweating, and the victim then suffers from convulsions and death. In the 1950s, British scientists developed an even more dangerous nerve agent—VX—which was subsequently weaponized by the United States. It kills in a similar way to sarin, but whereas sarin quickly evaporates, VX is more jelly-like and so can contaminate the environment for longer periods of time; the United States packed the chemical into land mines to render large areas inaccessible to the enemy.

In 1961, the Joint Chiefs of Staff authorized the head of the Pacific Command to store 16,000 tons of chemical weapons on Okinawa. In 1963, 11,000 tons of toxins were brought to Okinawa in two shipments and stored at Chibana Army Ammunition Depot. Another shipment arrived in 1965, bringing the total volume of chemical weapons to Okinawa to more than 13,000 tons. According to declassified US Army reports, these three shipments consisted of distilled mustard, sarin, and VX agents. They were bulk-sealed in one-ton drums and contained in ready-made weapons, including 13,000 VX landmines and almost 60,000 shells of sarin agent.[48] The stockpile held an apocalyptic volume. Experts estimate that one liter of VX agent is theoretically enough to kill thousands of people, meaning there was enough nerve agent on Okinawa to eliminate everyone in the region many, many times over.

US veterans and Okinawan base workers have described the conditions in which the chemical weapons were stored at the Red Hat site. The munitions were housed in bunkers dug into the hillsides, and the area was monitored by an animal early warning system of free-roaming goats and caged white rabbits, whose deaths, the military hoped, would signal any leaks. Soldiers patrolling the area were issued automatic syringes containing the nerve agent antidote atropine and ordered to inject themselves if they suspected they had been exposed. Small test kits containing samples of chemical weapons were also used in training sessions to help troops identify the substances.

According to the USAF, the Chibana depot's chemical weapons were flown off the island for testing in the United States. For example, in March 1969, twenty-five bombs, each containing one hundred kilograms of sarin,

*US soldiers inspect chemical weapon storage bunkers at Chibana Army Ammunition Depot in January 1971.* Okinawa Prefectural Archives

The US Army kept white rabbits at Chibana Army Ammunition Depot to monitor leaks of chemical weapons, January 1971. Okinawa Prefectural Archives

were driven from Chibana to Kadena Air Base by tractor trailers. Apparently, the route was not evacuated of civilians, nor were local authorities alerted to the dangers. Following the arrival of the munitions at Kadena, they were flown via Hawaii for outdoor trials at Dugway Proving Grounds in Utah.[49]

## OPERATION RED HAT

At the Red Hat site, the toxic stockpile required constant maintenance. Inside the weapons, the sarin was eating through the containers because the military had cut corners in its production, making the sarin acidic and laden with impurities. At the same time, Okinawa's humid, salty air was corroding the weapons' exteriors. According to the military, approximately thirty-two thousand munitions were defective.[50]

On July 8, 1969, Americans at the Chibana depot were sandblasting a 227-kilogram sarin bomb to prepare for repainting when it sprung a leak. Twenty-three service members and one US civilian were hospitalized, some of them for as long as a week. As soon as the accident occurred, the officer in charge of the chemical weapons telephoned the high commissioner of Okinawa and asked for permission to dump the leaky munition at sea; the commissioner told him he'd need permission from the Pentagon. The military did not make the accident public, but reporters from the *Wall Street Journal* soon discovered what had happened and their story hit headlines worldwide.[51]

Okinawans reacted to the news that chemical weapons were being stored in their midst with understandable fury, but again CIA officers were primarily concerned with the risks of negative publicity. "Japanese leftists . . . are hurting for a good rallying cause. . . . They may well be tempted to try to give this present incident a good propaganda ride," stated one report.[52]

Coming as it did in the middle of reversion negotiations, Washington policy-makers took the accident more seriously. President Richard Nixon was forced to announce the end of US chemical weapons production and pledge that the United States would only ever use existing stockpiles in retaliation. To Okinawans, the US government promised to remove its entire chemical weapon arsenal from the island. First, the military tried to ship them to Umatilla Chemical Depot in Oregon, but the plan was blocked by the governor; then the military decided to send them to Kodiak Island Naval Station in Alaska, but senators there thwarted the project. The military finally settled on the US territory of Johnston Atoll, an island under US control but outside of its laws. The island had been heavily contaminated with radiation from failed nuclear launches in the early 1960s.

Meanwhile, on Okinawa, the presence of the stockpile terrified nearby communities, and they held large demonstrations concerning the delays in removing it. On December 20, 1970, these contributed to the island's most violent anti-US riot in Koza City, where residents beat Americans, set alight their cars, and stormed Kadena Air Base; eighty vehicles were destroyed and sixty Americans hospitalized. The following month, removal work finally commenced, eighteen months after the initial leak.[53]

"Operation Red Hat," claimed the military, had been named by a "little old lady in tennis shoes working at the Pentagon," and it was accompanied by a cheery propaganda campaign replete with cute pin badges and a docu-

*In May 1970, Okinawans staged a demonstration calling for the removal of chemical weapons from their island and an end to US attacks on Cambodia.* Okinawa Prefectural Archives

mentary called "Operation Red Hat: Men and a Mission." According to the official version of events, it was a grand success. In January 1971, phase one of the operation involved soldiers driving nine trailer loads of chemical weapons from Chibana to Ten Gan Pier, where they were loaded onto ships. In August 1971, phase two of Operation Red Hat started. During a thirty-eight-day period, 1,213 trailer loads of munitions were loaded aboard. According to the Pentagon, the only mishap occurred when a pallet of sarin rockets was dropped into a ship's hold; despite being badly mangled, no leaks occurred.

As the final trucks passed by, Okinawan officials tossed handfuls of salt in a ritual to purify the area; finally, soldiers draped a large sheet over the ship's rail, painted with the *Loony Tunes* catchphrase, "That's All Folks!"

But it wasn't.

Veterans' accounts and scientific data suggest the US military has not been entirely truthful about what actually happened during Operation Red Hat.[54] The first mistruth, according to veterans, is how the military responded to the sarin leak at Chibana. They say the military followed the initial recommendation of the officer in charge and, as was standard operating procedure at the time, dumped the damaged munitions and others at sea. Former service members recall large volumes of chemical munitions driven by truck from Chibana to Tengan Pier in the autumn of 1969.

"It was an all-hands effort, with cooks, supply clerks, and road duty MPs manning intersections and escort duty. We all got the atropine training and were issued gas masks," recalled one MP.

Meanwhile, a US Army stevedore recalled the same convoy and said the chemical weapons had been packed into large steel containment tubes. Equipped with a gas mask and syringe of emergency antidote, he accompanied the weapons when they were transported from Tengan Pier to an unknown spot several hours sailing from Okinawa's coast.

"We used a winch to lift the containers out of the hold and then a large fork lift to push them overboard," the soldier further remembered. "I am not sure how many we dumped—it was a lot. After the first fifteen or twenty, I stopped counting. The entire operation took forty-eight hours."[55]

At the time, Okinawan newspaper journalists had heard the same accounts from on-base sources, but when reporters pushed for details, their informants suddenly clammed up for fear of losing their jobs.

Today, experts on chemical weapons disposal from the Monterey Institute in the United States estimate the metal containers for chemical weapons would have corroded to the point of rupture following approximately fifty years of submersion in sea water. In short, the munitions dumped off Okinawa are now approaching their safety limits.

According to the institute, sarin and VX nerve agents are believed to persist for long periods in ocean waters. Moreover, mustard agent, when exposed to sea water, develops a hardened outer shell that keeps its contents as toxic as when it was first manufactured.

Back on US-occupied Okinawa, two further leaks of chemical weapons reportedly occurred before the military finished removing its stockpiles. In December 1970, several Okinawan workers at Zukeyama Dam, near the Chibana depot, suffered injuries to their eyes and throats following a suspected leak within the installation. Another accident sickened three US service members in June 1971. One of them was Lindsay Peterson, an operations platoon leader for Operation Red Hat. While preparing a one-ton container of VX agent for shipment, a small amount leaked from around its rim. "One of the monitor rabbits died, and I began developing muscle twitches. They rushed three of us to the medical dispensary at Chibana and kept us there overnight for treatment," explained Peterson.[56]

Peterson also revealed how the US military disposed of chemical weapon test kits without informing the Japanese government. Each kit contained forty-eight small vials holding choking and blister agents. Peterson suspects the military either buried them on the island or sunk them at sea—but it seems they were dumped in a pond near the Chibana facilities; the Japanese government's 2003 chemical weapons survey noted the discovery of hundreds of "gas-filled containers" that the military refused to explain.[57]

The US military displayed similar negligence with its inventory of chemical weapons on Okinawa. In 1997, when scientists on Johnston Island belatedly conducted tests on the containers packed during Operation Red Hat, they discovered that a quarter of them had been mislabeled, and there were traces of one weapon—lewisite—not mentioned anywhere in the military's official accounts.[58]

According to US veterans, there was one more chemical the US military surreptitiously transported from Okinawa during Operation Red Hat: Agent Orange. One former US Army Special Forces Green Beret recalled seeing hundreds of barrels of defoliants loaded into the holds of one of the ships, and a former US soldier who helped offload chemical weapons on Johnston Island described similar barrels. "There were herbicides involved in Operation Red Hat. The barrels I sat on were fifty-five-gallon (208-liter) drums."[59]

Many of the Operation Red Hat veterans believe they were sickened from exposure to the chemicals involved in the mission, particularly Agent Orange, which the US government insists was never present on Okinawa.

# 4

# MILITARY HERBICIDES, VIETNAM, AND OKINAWA

During the 1920s and 1930s, scientists in Europe and the United States investigating ways to improve harvests found that the growth of plants was controlled by substances similar to hormones. When artificially synthesized and applied, these chemicals yielded better crops, but when used in excess volumes, they caused growth to spiral out of control, killing the plant. This destructive aspect caught the attention of the military, which realized these substances might enable it to destroy its enemies' crops and starve countries into surrender. In 1943, the US Army Chemical Corps created its Crops Division at Camp Detrick, Maryland, and during World War II, scientists researched thousands of different crop-killing substances; two that seemed to offer the most potential for military use were the herbicides, 2,4-dichlorophenoxyacetic acid (2,4-D) and 2,4,5-trichlorophenoxyacetic acid (2,4,5-T).[1]

By 1945, the US military was making plans to attack Japan with phosgene and mustard agent; it was also preparing to unleash its anticrop substances, dousing small Pacific islands with herbicides to force enemy troops into surrender and attacking rice crops on the homeland. The atomic bombings of Hiroshima and Nagasaki rendered these tactics moot, and following the war, Camp Detrick scientists received a boon from Unit 731 scientists' anticrop reports, which they said contained "much interesting and worthwhile information."[2]

During the Cold War, anticrop research, along with other biochemical weapons work, escalated at Camp Detrick. Eager to test its herbicides in actual combat, in 1952, the military shipped five thousand barrels of Agent Purple to Guam for use in Korea, but once again, the end of fighting thwarted the Pentagon's plans to use it. In the 1950s, the British military used 2,4,5-T in Malaya, and this encouraged the United States to continue its herbicide research under Project AGILE, whose remit also included other unconventional warfare techniques, for example, booby traps and psychological propaganda. Tests on different formulae and spray systems were conducted in the mainland United States and overseas.[3]

In 1961, scientists from recently renamed Fort Detrick brought biological weapon tests to Okinawa, where they experimented with rice blast at three different sites. According to a high-ranking US officer interviewed by the *Okinawa Times*, herbicide tests were also conducted between 1960 and 1962, in the northern Yanbaru jungles. "Within twenty-four hours of the spraying, the leaves had turned brown. By week four, all the leaves had fallen off. It was confirmed that weekly spraying stopped new buds from developing," the official said. The Department of Defense had chosen Okinawa, he explained, because its jungles resembled those in Southeast Asia where the United States was considering deployment of the chemicals and also because the military's jurisdiction over Okinawa enabled it to bypass more stringent safety standards that might limit such research elsewhere.[4]

## VIETNAM: NATION AS LABORATORY

Explaining the Pentagon's approach to Vietnam in the early days of the conflict, General Maxwell Taylor said, "On the military side . . . we have recognized the importance of the area as a laboratory." Many of the chemicals it began to employ in the region were new, for instance, deadlier versions of napalm that could keep burning flesh even under water, and more potent CS gas, of which it used millions of kilograms.[5]

The US military first dubbed its defoliation campaign Operation Hades, but public relations advisors quickly changed the title to Operation Ranch Hand, a name that, instead of conjuring up desolate visions of Greek hell, tapped into imagery of cowboys taming unruly nature. The logo chosen for their badges was *murasaki*, the Japanese kanji character for purple reflecting Agent Purple, the first formulae produced in bulk.

The United States has always contended that its use of herbicides did not breach international conventions against chemical weapons because the substances targeted plants, not humans; however, the Pentagon was initially so wary of international criticism that it launched Operation Ranch Hand in secrecy. Disguising its herbicides as civilian supplies, it flew them into Vietnam to bypass compulsory customs inspections of cargo ships. Moreover, the US ambassador to Vietnam recommended that United States Air Force (USAF) flight crews not wear military uniforms, and upon their arrival at Tan Son Nhut, near Saigon, the C-123 spray planes were stationed out of sight in an isolated corner of the base.[6]

In January 1962, the first official Operation Ranch Hand C-123 flight took off and sprayed herbicides along a road near Saigon to clear vegetation in the hope it would prevent guerrilla ambushes. This flight became the first of 19,905 defoliation sorties conducted in the next ten years. During this time, the United States sprayed approximately 76 million liters of experimental herbicides over South Vietnam and neighboring Cambodia and Laos.[7] Many of these missions targeted food crops—rice, maize, sweet potatoes, and bananas—to starve enemy troops and force South Vietnamese peasants off their land into zones where they might be better controlled by US-backed authorities. As well as being spread by aircraft, herbicides were also dispersed from helicopters, trucks, and boats; additionally, service members sprayed them by hand along perimeter fences to deter infiltration. Veterans have also recalled using the same herbicides in Guam, Okinawa, and Thailand to clear vegetation from the edge of runways and such important infrastructure as fuel pipelines and antenna farms.

Such heavy use left behind thousands of empty barrels, which US soldiers and Vietnamese civilians converted into makeshift showers, cut in half for barbecues, or used to store gasoline. The latter produced an unforeseen consequence: On average, the barrels contained two liters of herbicide residue, and the exhaust fumes of vehicles refueled from these drums stripped trees bare, including those of Saigon's famous boulevards.[8]

During the war, the US military worked with thirty-seven manufacturers, notably Dow Chemical Company and Monsanto Company, continually tweaking the formulae of its herbicides. To distinguish between different mixtures, they painted the barrels with a colored stripe or lid. These "rainbow" herbicides included Agent Blue, which was used to kill grasses; Agent White, which worked against woody vegetation; and Agent Orange, the most commonly used formula, effective against broadleaf jungle. In total, the military employed twelve types of herbicides during the Vietnam War, as well as chemicals to sterilize the soil and ensure vegetation would be not able to grow back. Agents Blue, White, and Orange were also available to order via military supply catalogs by individual US base commanders at a cost of $385 for a 208-liter barrel.[9]

The military's record-keeping of the procurement and use of its rainbow herbicides was lax. Following the war, 14 percent of the Agent Orange sent to Vietnam remained unaccounted, 400,000 liters of Agent Pink had been lost, and data on Ranch Hand spray missions were incomplete. Likewise, some of the records of herbicide shipments made prior to 1965 are missing.[10] The lack of documentation is even worse for herbicide missions in Cambodia and Laos, which were covertly run by the CIA.

Throughout Operation Ranch Hand, the United States and its puppet South Vietnamese governments staged a propaganda campaign extoling the safety of its herbicides via the media, which, at the time, was being tightly controlled. In 1962, for example, the press ran an announcement that the herbicides would not "harm wild life, domestic animals, human beings, or the soil." Leaflets were also dropped from aircraft into areas targeted for defoliation, promising farmers the chemicals were harmless. US military personnel were given similar assurances: "(Agent) Orange is relatively nontoxic to man or animals. No injuries have been reported to personnel exposed to aircraft spray."[11]

## THE HUMAN TOLL OF US HERBICIDES

Vietnamese soldiers, civilians, and American troops who experienced defoliation operations firsthand recall what struck them most following the wake of the C-123s was the silence. Jungles fell quiet as the chemicals killed first the insects, then small birds and frogs; sometimes monkeys fell stunned from the trees. In village farms, flocks of chickens and ducks died, and fish floated dead to the surface of ponds.

US defoliation missions sprayed an estimated 43 percent of South Vietnam's farmland and 44 percent of its forests at least once; the mangrove swamps of the Mekong Delta were devastated.[12] By 1965, almost half of C-123 flights targeted crops, destroying food supplies for a million people which led to mass starvation among civilians but had little impact on enemy troops.[13]

As many as 4.8 million Vietnamese people were exposed to military herbicides and an estimated 3 million fell ill.[14] The health impact depended on the chemical sprayed, the degree of exposure, and individual tolerance. For some, initial reactions included irritated skin, vomiting and diarrhea, headaches, and numbness of the fingers and toes. In the coming months and years, these symptoms deteriorated into diabetes, misfiring immune systems, and cancers. Doctors in the United States and Vietnam looked on

helplessly as men and women in their twenties and thirties developed illnesses more common in people double their age.

Then there were the babies. One study found that two-thirds of Vietnamese exposed to US herbicides had children suffering from serious
health problems.[15] They died in the womb or, in the words of one Vietnamese writer, were born as flapping, gasping fish that died almost immediately. Risking the wrath of the authorities, Saigon newspapers published
photographs of these babies—one with a "face like that of a duck," another
with a "head that resembles that of a poodle or a sheep," and another
with "2 heads, 3 arms, and 20 fingers."[16] Other children born apparently
healthy passed away after a week, a month, or a year. In South Vietnam,
the US-backed authorities blamed the deaths of these babies not on military herbicides, but on a fictional venereal disease they dubbed "Okinawa
bacteria."[17]

Despite the US military's repeated assurances that its herbicides were
safe, they were toxic. Agent Blue, the herbicide sprayed over rice fields,
contained an arsenic compound, cacodylic acid, which damages the fetus
and causes cancer. During the war, more than 4.7 million liters of Agent
Blue was used in Vietnam. Agent White contained two substances—HCB
and nitrocamines—both of which are known carcinogens. 2,4,5-T, the
ingredient in agents Pink, Purple, and Orange, was contaminated with
2,3,7,8-tetrachlorodibenzo para dioxin (TCDD), described by the World
Health Organization (WHO) as "highly toxic and can cause reproductive
and developmental problems, damage the immune system, interfere with
hormones, and also cause cancer."[18] Of the herbicides used during Operation Ranch Hand, 65 percent contained TCDD, and during the Vietnam
War, the total weight of dioxin sprayed by the US military was 366 kilograms; TCDD's toxicity is measured in the trillionths of a gram.[19]

The presence of dioxin in the herbicides was a result of the manufacturers' greed and the hunger of the military for more chemicals. When
herbicides are produced at low temperatures during a long period of time,
the amount of dioxin is relatively small; however, during the Vietnam War,
the military's demand was so high that the makers decided to cut corners to
meet orders, so they operated their factories around the clock.[20]

Perhaps one comment by General William Westmoreland, the man in
charge of the war, best encapsulates the Pentagon's disdain: "The Oriental
doesn't put the same high price on life as does the Westerner. Life is plenti-
ful, life is cheap in the Orient. As the philosophy of the Orient expresses it,
life is not important."[21]

## HALT TO OPERATION RANCH HAND ONLY
## THE START OF PROBLEMS

On April 15, 1970, the deputy secretary of defense, David Packard, declared an immediate suspension on the use of herbicides containing 2,4,5-T, particularly Agent Orange, in Vietnam. While Agent Blue could still be used and some herbicides were still sprayed by hand, truck, and helicopter, C-123 missions slowed to a halt. The Pentagon had been forced to act; the US media had obtained a scientific report citing incontestable proof of the dangers of these herbicides.

In 1966, Bionetics Research Laboratories had been working under contract to the US government to investigate the effects of 2,4,5-T on laboratory animals. It had discovered that, even in low doses, the substance caused birth defects. The laboratory had relayed its findings to the manufacturers and the US government, but for four more years, the military continued to spray its herbicides, and the producers continued to reap huge profits, aware of the risk to both US troops and the people of Vietnam. They probably would have kept spraying for many more years if not for a graduate student majoring in biology at Harvard who obtained a copy of the Bionetics report in late 1969, and leaked it to the press.[22]

However, the revelations about 2,4,5-T's toxicity failed to pave the way for help for those suffering from exposure to military herbicides. In 1978, the VA sent a secret memo to its branch clinics, ordering them not to award benefits to Vietnam War veterans suffering from exposure to herbicides. Veterans' claims were stalled or their paperwork was lost, or sick service members were sent on wild-goose chases to find records that never existed. Some troops' herbicide-related health problems were blamed on alcoholism, drug addiction, or venereal disease.[23]

In 1979, US veterans launched a class-action suit against the companies that had produced herbicides for the military. By 1984, it had become clear to the makers that they would almost definitely lose the case, so they offered to settle out of court. Without any input from the service members themselves, the veterans' lawyers agreed to a settlement of $180 million; those who had been totally disabled by their exposure received $12,000 paid during a span of ten years, while the lawyers pocketed $13 million.[24]

For the manufacturers, the lawsuit was an overwhelming victory. They made it clear that the one-off payment was in no way connected to an admission of guilt, and it protected them from future litigation. This hobbled other groups seeking compensation for their illnesses. In 2004, Vietnamese victims launched a similar lawsuit against the manufacturers, but it was rejected outright. Dow maintains that, "The very substantial body of human evidence on Agent Orange does not establish that veterans' illnesses are caused by Agent Orange," and Monsanto insists that, "A causal connection linking Agent Orange to chronic disease in humans has not been established."[25]

Throughout the 1980s, the US government launched several showcase research projects into the health effects of military herbicides. It did everything it could to tamper with the data to disprove links between exposure and illnesses. It skewed statistics, faked numbers, and buried crucial findings. For example, a 1984 report that showed increased health problems among Ranch Hand veterans was altered to hide the truth, as was a 1987 investigation that revealed highly elevated cancer risks for other exposed veterans. In 1990, Admiral E. R. Zumwalt Jr. summarized government studies as characterized by "deception, fraud, and political interference."[26]

One of the experts most active in US government campaigns to downplay the connection between military herbicides and health damage has been Dr. Alvin Young, a former USAF scientist. Since the 1970s, Young has published a series of reports denying that Agent Orange and other rainbow defoliants harmed those exposed. Working as a consultant for the Department of Defense and the VA, his studies have been used by the government to avoid or delay paying compensation to veterans; his work has also been funded by the makers of the herbicides, Monsanto and Dow.[27]

Young has argued that only a small number of service members came into contact with herbicides and such exposure was not harmful; he once called veterans claiming sicknesses "freeloaders." His fellow scientists have criticized the reliability of his work and close relationship with the government and herbicide manufacturers. Young has defended his research and contends there is no evidence that categorically proves military herbicides have caused any illnesses.[28]

US government refusals to recognize these illnesses have infuriated exposed veterans. After a long campaign for recognition, the Agent Orange Act of 1991 finally granted help to Vietnam War veterans suffering from diseases caused by exposure to military herbicides. Any veteran who served in Vietnam and suffers from a list of fourteen illnesses, some of which are prostate cancer, Hodgkin's disease, and type 2 diabetes, is automatically able to receive help. In addition, the VA also helps the children of veterans who suffer from a limited list of health problems, for example, spina bifida. By 2018, more than 750,000 veterans who had served in Vietnam had received benefits from the VA for exposure to herbicides.[29]

"Agent Orange" has become a catchall phrase for the herbicides used during the Vietnam War. Recognizing this scope, the VA uses the phrase "herbicide agent," which includes the following components: 2,4-D; 2,4,5-T and its contaminant, 2,3,7,8-TCDD; cacodylic acid; and picloram.[30] But the US government refuses to compensate the Vietnamese, Laotian, or Cambodian people exposed to its herbicides. In Vietnam alone, there are approximately 3 million people suffering from the effects of dioxin; these include those directly exposed, their children and grandchildren, and those exposed from the environment. This failure to help places an appalling burden on survivors, their families, communities, and the nation as a whole.[31]

Today, almost every community in Southern Vietnam has a care facility dedicated to those poisoned by US herbicides. Some victims sit without arms or legs, while the flesh of others has grown over where their eyes should be or their skin sheds at the slightest touch. Children lie in beds, their heads so swollen by hydrocephalus that they would topple if they tried to stand; others roll, groan, and bang their faces against the bars of their cots. At Tu Du Hospital, Ho Chi Minh City, one of the first places where the link between herbicides and birth defects was identified, more than a hundred deformed babies are preserved in formaldehyde jars bearing dates, their mothers' names, and the defects that killed them.

*In Southern Vietnam, millions of people are suffering from the effects of US herbicide use, one of whom is this young patient at Tu Du Hospital, Ho Chi Minh City. Photo by Jon Mitchell*

Many people in Vietnam think the United States ought to have been taken to the international war crimes court and punished for its chemical warfare. Since the 1960s, when the US government began to deny herbicides were a form of chemical weapon, it has repeatedly downplayed their impact in Vietnam. Unsurprisingly, Japan supported US assertions, arguing in 1970, that herbicides ought not to be added to the Geneva Protocol prohibiting chemical weapons. As recently as 2003, the US Embassy in Vietnam accused the Hanoi government of waging a "two-decade-long propaganda campaign" on military herbicides. Then the following year, the ambassador said Vietnamese complaints were based on "fake science."[32]

Dioxin is a persistent contaminant that, beneath the ground, has a half-life of as long as a century. Today in Vietnam, there are approximately thirty dioxin hot spots where the United States stored herbicides during the war. In the early 2000s, North American scientists identified several dioxin hot spots in and around Da Nang airport, one of the key Operation Ranch Hand bases. Dioxin-laden soil had accumulated at the bottom of ponds where residents kept fish and ducks, and grew lotus root; mothers had spread dioxin to their babies through breast milk. According to postwar surveys, some Da Nang residents had dioxin in their bodies more than a hundred times over safe levels.[33]

In the twenty-first century, the United States and Vietnam wanted to improve their diplomatic and trade relations, so the United States began to make token efforts to atone for Operation Ranch Hand. In 2012, Washington launched a cleanup of dioxin contamination at Da Nang Airport, insisting the project was solely environmental, with no acknowledgment of any health impact from dioxin. Remediation was completed in 2018. Then, in 2019, the United States and Vietnam announced they would commence cleanup at the former Bien Hoa Air Base, the country's largest dioxin-contaminated site.[34]

Although the United States has finally started to help Vietnam, it has extended no such support to Laos and Cambodia, where the US military and CIA drenched the nations during their secret wars. Likewise, for decades the United States refused to acknowledge its use of herbicides elsewhere in Asia. In 2010, the VA finally recognized they had been sprayed along the perimeters of bases in Thailand between 1961 and 1975; then, in 2011, it acknowledged their use along the Korean demilitarized zone between 1968 and 1971.

As of 2019, the Department of Defense has listed some fifty locations in the United States and twenty-one sites overseas where it admits to spraying or storing herbicides. The list has helped veterans receive support if they served in the locations and have been sick from exposure; however, veterans believe the Department of Defense has omitted many locations. In 2018, the Government Accountability Office (GAO) criticized the list as "inaccurate and incomplete," and complained that, "Neither DoD nor VA has taken steps to validate or correct the list." These failures have prevented veterans from receiving the help they deserve; as of June 2018, only approximately 11,000 veterans had been granted VA service connection for herbicide exposure outside Vietnam, while 58,000 had been denied and 23,000 were still awaiting decisions.[35] One of the sites where the Department of Defense still denies the presence of herbicides is the island without which it could not have fought the Vietnam War: Okinawa.

## MILITARY HERBICIDES ON OKINAWA: US VETERANS SPEAK OUT

Denials of herbicides on Okinawa by the United States date back to July 2004, when General Richard Myers, chairman of the Joint Chiefs of Staff, stated that government "records contain no information linking use or storage of Agent Orange or other herbicides in Okinawa." In April 2011, the *Japan Times* published my article "Evidence for Agent Orange on Okinawa," which featured the first accounts of service members who believed they had been exposed on the island. One recalled unloading hundreds of leaking barrels of Agent Orange at Naha Port, another remembered spraying the same substance around antennas and perimeter fences, and a third recalled an accident in which a soldier was drenched in the chemicals. All three men fell ill, and some of their children were sick, too.[36]

During the next six years, I wrote dozens more articles based on interviews of more than a hundred US veterans who claimed they had unloaded, stored, or sprayed military herbicides on Okinawa during the 1960s and 1970s. Scott Parton, a marine based at Camp Schwab, recalled seeing dozens of barrels of Agent Orange at the installation between 1970 and 1971. Many of them were leaking, and he remembered their contents being sprayed on the base; Parton had a photo of one of the striped barrels being used to burn trash on the shoreline, a dangerous practice given the two-liter residues they often contained.[37]

Another marine alleged she had been exposed at Camp Foster between 1975 and 1976. Caethe Goetz, pregnant at the time, said she often saw teams spraying herbicides along the perimeter, and once they blew onto her face. The daughter she was carrying developed spinal disorders, and she herself developed multiple myeloma, a rare form of cancer, at the age of forty-nine.[38]

Larry Carlson was a soldier assigned to the 44th Transportation Company at Naha Port between 1965 and 1967. He recalled helping offload barrels of herbicides, including Agent Orange, and dispatching them to bases on the island, one of which was Kadena Air Base. Carlson had contracted lung cancer and Parkinson's disease; two of his daughters had given birth to stillborn babies.[39]

Some of the veterans explained that the military preferred using herbicides to mechanical means because spraying did not set off World War II unexploded ordnance (UXO) or expose maintenance crews to habu snakes. Other veterans recalled selling defoliants destined for Vietnam to Okinawan farmers, who valued the chemicals' jungle-killing strength.

Several service members recalled the burial of empty and damaged barrels of defoliants on Okinawan bases. For instance, one former soldier stationed on the island between 1968 and 1970 claimed he had taken part in the burial of dozens of barrels of Agent Orange that had been damaged in a shipping accident. The burial, he claims, took place at Hamby Yard, a facility adjacent to an airstrip run by the CIA's Air America. Such disposal techniques were

Former US soldier Larry Carlson believes he was exposed to military herbicides while unloading them at Naha Port; here at his Florida home he shows the scar from his lung cancer operation. Photo by Jon Mitchell

standard operating procedure for the military; one army training manual stated, "Used containers and surplus quantities of ORANGE should be buried in deep pits at locations where there will be the least possibility of agent leaching into water supplies or cultivated crop areas."[40]

Piecing together the testimonies from these veterans suggested that military herbicides, including agents Orange and Pink, had been stored, sprayed, or buried within at least fifteen facilities on Okinawa. Some of the veterans had photographs of barrels of Agent Orange to support their allegations, while others had signed buddy statements from fellow service members corroborating their claims. They also had documentary proof. In 1971, responding to the US ban, officials from Fort Detrick compiled an inventory of sites where defoliants had been stored in a report titled "Historical, Logistical, Political, and Technical Aspects of the Herbicide/Defoliant Program." In it, they cited a "herbicide stockpile" at Kadena Air Base, as well as in Thailand, which is notable given the Department of Defense's later inclusion of the country on its list of storage sites. Another report, "An Ecological Assessment of Johnston Atoll" (2003), funded by the US Army Chemical Materials Activity (CMA), stated, "In 1972, the US Air Force brought about twenty-five thousand 55-gallon (208-liter) drums of the chemical Herbicide Orange (HO) to Johnston Island that originated from Vietnam and was stored on Okinawa."[41]

## THE IMPACT ON OKINAWAN CIVILIANS

The concentration of military facilities on Okinawa also concentrated the contamination; US herbicides harmed civilians. According to a report from Okinawa Prefectural Archives, in December 1961, farmers in the Yanbaru witnessed a "Nisei employee of the American Chemical Company" spray herbicides from a truck to clear weeds along roads leading to radio antennas. Two cows ate the weeds that had been sprayed and, the following day, developed symptoms that included "bloody diarrhea," increased salivation, and dehydration. One of the cows died, while the other was euthanized to end its suffering.[42] According to the report, the ingredients of the herbicides included 124 kilograms of arsenic.

Throughout the mid- to late 1960s, Okinawan newspapers carried reports about suspected damage from military herbicides, for instance, explaining how farmers' green peppers, papaya, and goya were found with deformities in the Chibana area. In response to such coverage, military authorities either denied links to its operations or refused to comment.

In July 1968, young Okinawans apparently fell victim to military herbicides. One day, a group of students from a school in Naha were swimming in the sea in Gushikawa. Soon after entering the water, they started complaining that their skin felt as if it were burning. Rushed to hospital with swollen lips, sore eyes, and rashes, some of them were burned so badly they had to stay as in-patients. More than 230 children were injured. The local media suspected the incident had been caused by defoliants, but the US military refused to confirm it.

More than forty years later, Okinawan TV director Shimabukuro Natsuko and I tracked down the apparent proof. We interviewed one of the American service members who'd been spraying the stretch of shoreline with defoliants when the children were burned. His account was corroborated by an Okinawan base worker who had seen the orange-striped barrels along the shoreline. During the same period, newspapers had also reported the discovery of dozens of deformed frogs in nearby rice fields.[43]

A CIA memo from 1969 reveals a familiar attitude, more concerned with public image than human health: The Gushikawa incident "blew over without serious repercussions, largely due to the relative mildness of the affliction and the inability to pinpoint the cause."[44]

## IEJIMA ISLAND

In 1955, as part of the Pentagon's goal to further militarize Okinawa, it decided to build an air-to-surface bombing range on Iejima Island. First, the military attempted to trick farmers off their property by asking them to sign voluntary eviction notices written in English, a language they couldn't understand. When some villagers refused to relocate, the military moved in with bayonets and bulldozers. They dragged sick children and old people from beds, and flattened their houses. At night, US soldiers got drunk on looted *awamori* liquor and shot the farmers' goats.[45]

In response, Ahagon Shoko, an Iejima resident, led the dispossessed islanders on a seven-month march on mainland Okinawa to inform people of what had happened. The march proved to be the start of the Okinawan peace movement, and throughout the 1950s and 1960s, under the guidance of this "Gandhi of Okinawa," Iejima farmers defied US orders not to grow crops on the outskirts of the bombing range. This persistence infuriated the military, which waged a battle of attrition against them, using gasoline to raze their fields. Tensions escalated in 1966, when Iejima's farmers successfully blocked the deployment of Hawk missiles to the island.

Against this acrimonious backdrop, in October 1973, US forces resorted to a new tactic. According to the *Okinawa Times*, the military sprayed defoliants in an area measuring two thousand square meters. Villagers lost their pasture land and worried about pollution of the nearby shore and the effects on their health. They filed a complaint with the US military, but whether the military bothered to respond is not known.[46]

The US military's use of defoliants on Iejima—the birthplace of the Okinawan civil rights movement—reveals a sickening degree of brutality. In 1971, the White House had banned these substances in Vietnam, but on Iejima, two years later, it was apparently employing them against peaceful demonstrators; the incident must surely warrant a full enquiry by Tokyo and Washington.

## CONTAMINATION AT MCAS FUTENMA, THE WORLD'S MOST DANGEROUS BASE

US veterans claim Marine Corps Air Station (MCAS) Futenma possessed a stockpile of rainbow herbicides both during and after the Vietnam War. Carlos Garay, a marine with the Headquarters and Maintenance Squadron at Futenma in 1975, claims he saw twelve barrels of Agent Orange at the installation. "Additionally, other squadrons were directing their leftover stocks to us for disposal, so I sent messages to the Department of Defense and Headquarters of the Marine Corps, but they never replied. The barrels were still there when I left in 1976," he said. Five years later, a larger cache was unearthed on the base.

In 1981, Lieutenant Colonel Kris Roberts was head of maintenance at MCAS Futenma. One day, his superior officers informed him they had a problem: The waste rainwater flowing into civilian areas was displaying dangerously high chemical readings. Roberts was ordered to remedy the situation. Digging in the problem area, Roberts and his crew of American and Okinawan laborers discovered a buried cache of approximately one hundred barrels—some with orange stripes around the middle. Roberts's superior officers declared the area off-limits to other personnel and ordered Okinawan workers to load the barrels onto trucks and transport them to an unknown location off the base. Roberts was suspicious about the response, so he snapped some Polaroid photographs of the scene, one of which shows young marines lifting the drums from a deep hole without wearing safety equipment or even shirts.

Following the removal, a storm hit the site. "It threatened to flood the runway so my crew and I climbed into the water to open the release gates," Roberts remembered. "The water had a chemical film on it from the leaking barrels. Eventually, we managed to drain the contaminated water off the base."

As a result, Roberts, a former gold medal marathon runner, developed serious illnesses, including heart disease and prostate cancer. Roberts regrets how Okinawan workers were used. "Those men were easily replaced. So if we told an Okinawan worker to do something, they did it. It wasn't fair."[47]

Further concerns about the impact of herbicides on Okinawans followed a lecture I gave in November 2011, near Camp Schwab. Residents explained how they believed the runoff from herbicide usage had contributed to the

*US marines unearth barrels of chemicals—including suspected herbicides—from MCAS Futenma in the summer of 1981.* Photo by Kris E. Roberts

destruction of the local *mozuku* seaweed industry and worried defoliants had sickened people who'd eaten shellfish collected in the area, particularly residents who had died soon thereafter. Okinawan TV ran a special report about the lecture, and several years later, I discovered via a Freedom of Information Act (FOIA) request that the CIA had created a full translation of its Japanese contents, replete with screenshots and a six-minute video file. My work, it seems, was reaching the eyes and ears of those who mattered.[48]

## THE OFFICIAL REACTION

The research I conducted from 2011 onward made frequent headlines in Japan and formed the basis of several award-winning TV documentaries. Public concern pushed Okinawan mayors to demand that the Japanese government investigate the matter, and even Okinawa's conservative governor, Nakaima Hirokazu, requested that US ambassador John Victor Roos look into the veterans' allegations.

The US military issued blanket denials. "In response to the Embassy of Japan request on August 10 (2011), DOD has once again searched and once again been unable to locate any record of Herbicide Orange or its component ingredients being used in Okinawa," said one US Forces Japan spokesman.[49]

Another US official told Tokyo, "There are some elements in the veterans' accounts which are questionable." With regards to the CMA-funded report citing twenty-five thousand barrels on Okinawa, the military told the Japanese government, "The description that '[Herbicide Orange] was stored in Okinawa' is inaccurate and contradicts with the facts that the US government acknowledges."[50]

Then, in February 2013, Department of Defense officials, members of the State Department, the VA, and representatives of the Japanese embassy met in Washington to hear from the man the United States hoped would, once and for all, disprove that military defoliants had ever been on Okinawa: Dr. Alvin Young. Under Department of Defense contract, Young had penned a twenty-nine-page report entitled "Investigations into the Allegations of Herbicide Orange on Okinawa," which boiled down my articles and veterans' testimonies into seven bite-size chunks and dismissed them one by one. Concerning the claim of herbicides arriving on Okinawa by ship, for example, Dr. Young wrote, "There were no records found." Countering Kris Roberts's account of the burial of Agent Orange on MCAS Futenma,

Dr. Young rejected it, writing, "There were no records or other evidence." The report concluded,

> After an extensive search of all known and available records, there were no documents found that validated the allegations that Herbicide Orange was involved in any of these events, nor were there records to validate that Herbicide Orange was shipped to or through, unloaded, used, or buried on Okinawa.[51]

The report infuriated veterans and Okinawan residents, none of whom had been interviewed for it, nor had any environmental tests been conducted on Okinawa where service members claimed the herbicides had been stored. The Pentagon's report brought full circle its fifty-year history of Agent Orange denials. But sometimes the truth can reveal itself in the most unlikely places.

## THE OKINAWA CITY SOCCER GROUND

In June 2013, four months after Dr. Alvin Young delivered his Okinawa report, laborers were working to install a sprinkler system beneath a children's soccer field in Okinawa City. Located just outside Kadena Air Base, the land was once part of the installation itself but had been returned to civilian control in 1987. While digging beneath the pitch, workers discovered rusty barrels, some stenciled with the Dow logo. In the following months, they unearthed 108 drums.

Tests revealed the barrels contained the three main ingredients of military defoliants: 2,4,5-T; 2,4-D; and TCDD. Nearby water was contaminated with dioxin at levels twenty-one thousand times the safe standard. The barrels also contained polychlorinated biphenyls (PCBs), pentachlorophenol, and arsenic. The hazardous solvent dichloromethane was discovered at 455,000 times the safe level.[52]

The US military's reaction to the discovery of the barrels was predictable. At first it tried to deny the barrels had been theirs. Dr. Young was quick to dismiss the results, suggesting that the barrels might have contained "waste from military hospitals and dining facilities." In a meeting, the head of Kadena Air Base likened the barrels to empty cans of tomato sauce, and the base produced a fact sheet assuring service members that dioxin caused the skin disease chloracne but "no other human health effects have been proven." This contradicts the findings of the Environmental Protection

A construction crew unearths a chemical barrel on an Okinawa City soccer pitch in January 2014.
Photo by Kuwae Naoya

Agency, which has stated that dioxin "can cause cancer, reproductive and
developmental problems, damage to the immune system, and can interfere
with hormones."[53]

Finally, the Okinawa Defense Bureau argued that because the barrels
did not contain the herbicides in equal measure, it couldn't possibly be
Agent *Orange*. It was a semantic sleight of hand; they said nothing of the
other herbicides. And the smokescreen flew in the face of the VA's official
definition of "herbicide agent" as a chemical that includes 2,4-D; 2,4,5-T;
and TCDD—all of which were found in the barrels.[54]

Experts agreed the barrels were the smoking gun. Wayne Dwernychuk,
a Canadian scientist who had spent fifteen years leading investigations into
dioxin hot spots in Vietnam, stated,

> The inescapable fact is that the US military, on land then part of Kadena Air
> Base on Okinawa, disposed of "unknown" materials in drums containing 2,4,5-T,
> a wartime herbicide/defoliant, and the most toxic component of the dioxin fam-
> ily, TCDD, known to be associated with the manufacture of such herbicides.[55]

Furthermore, in August 2013, Honda Katsuhisa, an Ehime University professor specializing in defoliants, stated the pattern of contamination closely

resembled the fields he'd previously researched in Vietnam.[56] But the Japanese government did not dispute the US military's conclusions, and the Department of Defense has not added Okinawa to its list of places where herbicides have been stored.

The Department of Defense cover-up and the Japanese government's collusion affected more than US veterans; countless Okinawan children had played on the soccer pitch, and the contamination had threatened the health of American children as well. Located adjacent to the dumpsite on Kadena Air Base were the Bob Hope Primary School and the Amelia Earhart Intermediate School, but the military had not notified parents of the discovery. Parents only found out six months later by chance, after a newspaper article was printed on the subject.[57]

Speaking on condition of anonymity because they feared for their military careers, parents described severe illnesses among their children, including cancers and autoimmune, respiratory, and neurological problems. All of them had attended the two schools or played on their fields between 1999 and 2013.

One of the parents brave enough to speak on record was Telisha Simmons. Simmons and her family were stationed at Kadena Air Base between 2011 and 2012; before arriving on Okinawa, none of them had experienced any serious medical problems. But during their time on the island, one of her sons developed a brain cyst and her daughter bone tumors; Simmons herself was diagnosed with a pituitary tumor and other serious illnesses, resulting in a hysterectomy at the age of thirty-five.

Simmons's children had attended one of the schools located near the dioxin dumpsite and regularly played on its fields. But the military has never investigated the family's health problems or the illnesses of the other children. "Kadena officials have known about this contamination the entire time, but they will do whatever they can to keep it all hush-hush," Simmons says.[58]

## HARD WORK PAYS OFF

According to the US military, prior to 2012, only two veterans had received help from the VA for their exposure to herbicides on Okinawa. One of them was a United States Marine Corps (USMC) driver who "reported that he had been exposed to Agent Orange while in the process of transport, as well as when it was used in Northern Okinawa for War Games training"; the veteran also stated that military defoliants were used "particularly near base camp perimeters. Spraying from both truck and back pack were utilized along roadways too." The marine had developed prostate cancer, which, in 1998, the VA ruled was a result of his exposure on Okinawa. The other was a marine stationed on Okinawa between 1972 and 1973, who had handled retrograde supplies from Vietnam contaminated with herbicides; in 2008, the Board of Veterans' Appeals (BVA) ruled the veteran's Hodgkin's disease and type 2 diabetes mellitus was a result of this exposure.[59]

However, the newspaper articles I'd written and the discovery of the 108 barrels paved the way for more sick veterans to receive the support they deserve. Since October 2013, at least nine more service members have been granted help from the VA for exposure to herbicides on Okinawa. One marine who had been exposed to Agent Orange on the island between 1967 and 1968 was awarded help for prostate cancer. According to the 2013 VA documents, he had come into contact with Agent Orange while transporting it between the island's ports and a warehouse located on Kadena Air Base. Another soldier stationed on Okinawa between 1972 and 1973 developed lung cancer, which, the BVA judged in 2017, had been triggered by his exposure to herbicides on Okinawa; the ruling was too late to help the veteran who passed away in 2011.[60]

In 2015, the US government finally awarded Kris Roberts compensation for his exposure at MCAS Futenma, although the decision did not cite such rainbow herbicides as Agent Orange, only *chemical exposure.*

## OKINAWA AGENT ORANGE WINS

How many service members were sickened by herbicides on Okinawa? In response to a FOIA request, the Veterans Benefits Administration says it "does not track claims for Agent Orange exposure based on Okinawa service"; however, a search of the BVA's publicly accessible database of rulings shows that, as of 2019, at least 250 service members had filed claims for compensation for exposure to Agent Orange on Okinawa. The actual number is far higher because the database only lists cases initially denied by the VA, appealed by the veteran, and given a final ruling. What is not known is how many veterans have applied for help, how many were awarded benefits, and how many decided not to appeal their denials or died before they could do so.

Veterans' hopes for justice were given another boost in 2018, by the GAO, which, during its research on herbicides on Guam, discovered records of at least two ships carrying Agent Orange that had docked in Japan,

**Table 4.1. Board of Veterans' Appeals Wins for Herbicide Exposure on Okinawa**

| Citation Nr./Decision Date | Location | Date Exposed | Illnesses |
|---|---|---|---|
| 18104750/May 23, 2018 | Kadena Air Base and Kwang Ju, South Korea | 1966–1970 | ischemic heart disease |
| 1804418/January 22, 2018 | "Okinawa" | 1968–1969 | liposarcoma |
| 1802686/January 11, 2018 | "Okinawa" | 1969–1972 | diabetes mellitus, type 2 |
| 1731591/August 7, 2017 | "Okinawa" | 1972–1973 | lung cancer |
| 1721360/June 13, 2017 | Kadena Air Base | 1965–1966 | coronary artery disease and diabetes mellitus, type 2 |
| 1635277/September 8, 2016 | Naha Port | 1968–1970 | diabetes mellitus, type 2 |
| 1543352/October 8, 2015 | Naha Air Base and Larson Air Force Base, Washington | 1954–1958 | prostate cancer |
| 1516681/April 17, 2015 | "Okinawa" and Fort A. P. Hill, Virginia | 1975–1979 | multiple myeloma |
| 1332861/October 21, 2013 | White Beach, Naha Port, and Kadena Air Base | 1967–1968 | prostate cancer |
| 0831082/September 12, 2008 | "Okinawa" | 1972–1973 | Hodgkin's disease and diabetes mellitus, type 2 |
| 9800877/January 13, 1998 | "Northern Okinawa" | 1961–1962 | prostate cancer |

contradicting Alvin Young's report stating there were no records to show it had been shipped to or through Okinawa.[61] Although US veterans are slowly receiving justice, there has been no such help for Okinawans, and the Japanese government has done nothing to help them. During the Vietnam War, fifty thousand Okinawans worked on the bases, but they have not been surveyed for health problems, nor have the farmers of Iejima or the residents living near Camp Schwab, MCAS Futenma, or the soccer field dump site.

Cleaning up the soccer pitch took months and showed appalling disregard for public safety, with no warning signs posted and many workers operating without protective clothing. After the 108 barrels and contaminated soil were removed, the area was covered with concrete and turned into a car park, at a cost of 979,000,000 yen ($8.9 million). The US government paid nothing—the entire bill was footed by Japanese taxpayers.[62]

*The soccer pitch in Okinawa City where the barrels of herbicides were discovered in June 2013; laborers work without safety gear after the site was inundated by a typhoon.* Photo by Ken Nakamura-Huber

# 5

# POLLUTING WITH IMPUNITY

The US military is the largest polluter on the planet. Globally, it consumes the most petroleum, making it the worst producer of greenhouse gases; between 2001 and 2017, it pumped out 1.2 billion metric tons of $CO_2$, more than some entire countries. Each year it creates more than one ton per minute of hazardous waste, exceeding that of the top three US chemical manufacturers combined. Between 2010 and 2014, it pumped almost thirty thousand tons of pollution into domestic waterways, a record beaten only by a steel conglomerate and a meat consortium. In oceans already polluted with civilian waste plastic, US Navy drills drop trillions of pieces of microfibers fired as radar-evading chaff.[1]

At the root of the problem is the sheer scale of the present US military. In the United States there are about 4,100 bases, and overseas there are approximately 800 facilities in more than 70 countries. Each of these facilities leaves its mark on the environment. Stateside, the military and its contractors have contaminated approximately 40,000 sites, ranging from large air fields and training zones to midsize missile bases and small warehouses; they total 162,000 square kilometers, an area almost double the size of South Carolina. In 2017, 900 of these areas were so toxic they were categorized as Superfund sites, the Environmental Protection Agency's (EPA) most serious level, reserved for zones requiring urgent attention.[2]

In 1993, a senior Department of Defense official concluded that military bases are "laced with almost every imaginable contaminant—toxic and hazardous wastes, fuels, solvents, and unexploded ordnance." Another stated that the cleanup of this contamination was the military's "largest challenge." Pollution escapes perimeter fences, leaking into the sea and local rivers; it also seeps into wells and groundwater. In the United States, contaminated sites can be found in every state, making military contamination a problem that touches the lives of all Americans.[3]

## A TOUR OF CONTAMINATED BASES

A whirlwind tour of the United States reveals some of the ways in which the military has polluted the nation. Starting on the East Coast, the United States Marine Corps' (USMC) Camp Lejeune lies in North Carolina. Established in 1941, the base contains shooting ranges, forest training areas, and residential districts. Camp Lejeune used to be the pride of the USMC, but it has become synonymous with contamination. Between the 1950s and 1980s, fuel and solvents leaked into on-base wells, polluting the drinking water at levels thousands of times higher than safe standards. Underground storage tanks seeped cancer-causing benzenes, and nearby laundries spilled dry-cleaning chemicals. The USMC was aware of the contamination but allowed service members and their families to continue drinking the water; when the government began to investigate, base officials lied to them.

As a result of the contamination, thousands of service members and their families have fallen sick with such illnesses as leukemia, breast and bladder cancers, and miscarriages. An estimated 900,000 people were exposed in one of the largest cases of drinking water contamination in US history.[4]

Traveling south, we come to Eglin Air Force Base in Florida. During the 1960s, the military sprayed dioxin-tainted herbicides there in tests designed to improve their use in Vietnam. Barrels of defoliants were also buried and still poison the land today. Elsewhere at the base, pollution from insecticides, depleted uranium, and arsenic has seeped into the ponds and rivers where residents go fishing.[5]

Heading west from Florida, we pass by the hundreds of polluted sites in the Deep South until we reach Colorado, home of the US Army's Rocky Mountain Arsenal. Chosen because of its inland location, far from both coasts, which put it out of range of enemy bombers in World War II, the installation manufactured chemical weapons, as well as napalm and white phosphorus. At the base, waste from decades of production was dumped into large outdoor basins, which then bled into the local drinking water supplies. In the 1980s, the area was dubbed the "most toxic mile on the planet."[6]

Moving west from Colorado, let's make a brief stop to understand how the military has poisoned California. At Edwards Air Force Base, there are about 470 suspected toxic hot spots. Much of this contamination, like that at other bases, has arisen due to daily operations. The solvents and degreasing agents, many of which are carcinogenic, used for cleaning aircraft engines run into the ground; firefighting exercises contaminate water sources with toxic foam; and storage tanks leak diesel and fuel.[7] The pollution has seeped into the groundwater, creating widespread plumes of chemical contamination. Also in California lies George Air Force Base, where, until its closure in 1992, women were warned not to become pregnant due to illnesses linked to contamination.[8]

Crossing the Pacific Ocean, we arrive at Hawaii, home to approximately 142 military properties and multiple polluted sites. At its Pearl Harbor Naval Complex, contamination has emanated from underground fuel tanks, dry cleaners, and spills from electrical transformers containing cancer-causing polychlorinated biphenyls (PCBs). In 2014, one hundred thousand liters of jet fuel spilled from the navy's storage tanks at Red Hill, Honolulu. On Kaneohe Marine Corps Base, more than a thousand homes were built on land polluted with high levels of pesticides; residents had not been warned of the dangers. In the military, families are housed in moldy, lead-painted, radon-emitting properties that if used for civilian tenants, landlords would be punished for multiple health code violations.[9]

## DECADES OF MALPRACTICE

The disposal of hazardous waste and surplus stocks of chemicals is one of the main factors contributing to the contamination at bases in the United States; for decades, the military got rid of them through burning, dumping, and selling them to the public. Throughout the Cold War, facilities maintained burn pits and incinerators where waste was set alight without any care for the service members involved in the operations or local communities. The military continues to burn munitions and explosive byproducts at approximately fifty sites, contaminating nearby areas with such toxins as carcinogenic heavy metals and perchlorate, which can disrupt thyroid functions. Another technique favored by the military was to pour toxic chemicals into outdoor ponds, where it was hoped the substances would naturally break down over time. Today, the folly of this approach is clear, as such practices often make the toxins more concentrated or mix them into deadlier combinations.[10]

A similar misguided faith in nature's ability to deal with waste was behind the military's dumping of munitions, radioactive materials, and chemical weapons into the sea. Between 1944 and 1970, the US Army disposed of 29 million kilograms of mustard and nerve agents, and 454 tons of radioactive waste into the ocean. In one of the codenames beloved by the Pentagon, Operation CHASE (Cut Holes and Sink 'Em) involved packing ships with conventional and chemical weapons, sailing them out to sea, and scuttling them in deep waters. Experts assumed such methods would render chemical weapons inert, but as Japanese fishing crews near Choushi City can testify, they were mistaken; these munitions have contaminated marine life and injured fishing crews whose nets have accidentally hooked them.[11]

Perhaps the most disconcerting way of disposing of hazardous substances was via public auction. In a program known as the Defense Reutilization and Marketing Service, the military sold unwanted chemicals to civilian companies and individuals, often without ensuring they were licensed to handle them. In the 1980s, there were instances of buyers purchasing military acids, pesticides, and solvents, and then dumping them; the military even added barrels of toxic waste to mixed auction lots so buyers were unaware of what they were receiving.[12]

While it is true that some of the military's lax disposal techniques resembled those of the twentieth-century civilian industrial sector, two key differences must be remembered. First, some of the substances dumped by the military included chemicals designed for the sole purpose of killing. One gram of VX agent, for example, has the potential to kill one thousand people, and the depleted uranium used in armor-piercing shells has a half-life of many million years. Second, as a government organization, the military has no profit margins to worry about or shareholders to hold it accountable. Consequently, the military was not motivated to reduce waste, and it dumped surplus chemicals at will. Moreover, unlike with civilian factories, neighboring communities had no right to demand inspections of bases suspected of dangerous environmental practices, nor was the military required to keep paperwork detailing where toxic substances were stored, poured, or buried.

## HIGH-PROFILE MILITARY ACCIDENTS AND GROWING ENVIRONMENTAL AWARENESS

Due to the secrecy surrounding many of the military's operations, for decades it was able to hide its environmental impact from public view; however, in the late 1960s, a series of high-profile incidents opened people's eyes. Dugway Proving Ground in Utah was established in 1942, to develop chemical and biological weapons; throughout the Cold War, thousands of top-secret outdoor tests were conducted there. In March 1968, one of these experiments involved a jet plane spraying VX agent on an uninhabited part of the base. The test was designed to see how the toxin dispersed in air, but strong winds and a malfunction in the spray system carried the agent further than expected.

In the following days, more than six thousand sheep died in the vicinity. As the story made news headlines, the military denied any responsibility, claiming the deaths of the sheep were not connected to the nerve agent accident. But as the truth of the situation came to light, the military was eventually forced to pay compensation to the farmers whose livestock had been killed.

In 1969, the military's nerve agent program was again in the news due to the leak at Chibana Army Ammunition Depot in Okinawa. The accident revealed, for the first time, that the United States was indeed storing chemical weapons overseas and the publicity forced the US government to renounce its first-use policy on chemical weapons and restrict research into biological agents to defensive purposes only.

More proof of the toxic dangers of US weapons came in April 1970, when Washington banned the spraying of its most widely used wartime herbicide, Agent Orange. Publicly, the military had long asserted that the substance was safe, but when scientific studies proving its dangers were leaked to the public, the government finally banned it—much to the anger of the Pentagon.

The incidents at Dugway and Okinawa, followed by the revelations about Agent Orange's toxicity, coincided with a time of burgeoning environmental awareness in the United States. In 1962, Rachel Carson's *Silent Spring*, which would go on to sell more than 2 million copies, warned how uncontrolled use of pesticides caused them to enter the environment, threatening human health. One of the chemicals was dichlorodiphenyltrichloroethane (DDT), a substance linked to liver damage, miscarriages, and cancers. As a result of the book's publication, awareness of environmental issues grew, giving birth to numerous citizens' organizations. In 1970, for example, the first Earth Day was established, to highlight the importance of protecting the environment.

In response to mounting evidence of humankind's damage to the planet and political pressure from an increasingly green public, the US government introduced new measures. In 1970, it created the EPA to safeguard the environment and public health, followed by the 1970 Clean Air Act, the

1972 Clean Water Act, and bans on such harmful substances as DDT. In the 1970s, presidents Richard Nixon and Jimmy Carter signed executive orders that applied environmental laws to all federal facilities, including military installations. These were groundbreaking moves by the US government to make the Department of Defense more environmentally responsible.

In the following years, the work to hold the military to account was taken up by numerous groups, including state legislators, civic organizations, and the media. Although progress was often slow, they were gradually able to enact improvements. The EPA shut down some military operations that harmed the environment and health of local communities.[13]

Judges began to hold military employees and service members responsible for environmental negligence. In 1989, three engineers were found guilty of multiple charges, one of which was the illegal dumping of toxic waste into the sewer system in the chemical weapons factory at Aberdeen Proving Grounds in Maryland. Other service members were also prosecuted for falsifying environmental reports and illegally disposing of hazardous chemicals.[14]

As time went by, the military was made to follow some of the same guidelines as civilian industries. It had to keep logs of environmental accidents on its bases, maintain inventories of hazardous waste, and dispose of hazardous waste in safer ways. Installations were also forced to treat contaminated water and install oil–water separators to reduce the pollution of local waterways.

In the United States, despite military pushback, the situation has improved, thanks to the introduction of a series of environmental programs. According to the Comprehensive Environmental Response, Compensation, and Liability Act, a system has been established to deal with the cleanup of military land. The Installation Restoration Program deals with hazardous substances on active bases, and the Military Munitions Response Program cleans up unexploded ordnance (UXO). Closed bases are handled by the Formerly Used Defense Sites Program, and, at least in theory, the military engages with local communities to update them on progress and plan for future land use.

The biggest victory has been a measure of transparency. Americans today have a much better picture of the damage its military has caused the nation. The EPA, for instance, maintains extensive data related to pollution on both active and closed military installations. This information is available online, where reports list contaminants and potential pathways to exposure, and provide contact information for those seeking more information. Such transparency enables service members and civilians to take precautions to limit their exposure. Neighboring communities can request that authorities connect their homes to drinking water sources away from potentially contaminated well water; likewise, they can avoid farming or fishing in certain areas.

Increased transparency also helps medical professionals understand the causes of illnesses in those exposed. Moreover, it enables some veterans to claim free health care in a country notorious for its exorbitant medical costs. Among the exposure for which sick service members can receive support from the VA are diseases linked to asbestos, radiation, defoliants, mustard agent, and toxins from burn pits. Although accurate figures do not exist, one estimate places the number of service members exposed to contamination at 4.5 million.[15] On paper, the improvements enacted by the federal government seem entirely rational. But at almost every stage, the military fought tooth and nail to avoid such civilian oversight.

## THE MILITARY MINDSET

To understand the military's objections to becoming more environmentally responsible, it is necessary to understand its mentality. In his 2007 book *The Greening of the US Military*, Robert Durant sums up its deep-rooted attitude with the three Ss: "sovereignty, secrecy, and sinecure." In my own words, sovereignty means the military believes it should be allowed to operate independently of civilian influence—including that of the government and the citizens whose taxes fund it. Secrecy refers to issues of both national security and the right not to reveal how its annual budget—$700 billion for 2018—is spent. Sinecure reflects the military's sense of entitlement to its privileged status; it believes it is the defender of the Free World and so should be treated as exceptional.[16]

In a nutshell, the US military is a vast organization with an operating budget equal to the combined defense expenditures of the next seven highest-spending countries. The US military believes it should be allowed to operate outside the law and, as a result, blocks all civilian interference—including from its own government. Backed by a powerful lobby of arms manufacturers and bought politicians, today the US military is responsible to no one. As such, the military has perceived any attempts to control its environmental practices as a threat. Rather than seeing new laws as protecting service members and host communities from contamination, the Pentagon saw them as unwanted interference. As many in the Department of Defense were wont to quip, "The military is here to protect the nation, not the environment."

When national and local governments, citizens' groups, and the media began holding the military accountable for environmental damage, the Pentagon set about blocking them with the ruthlessness of a martial campaign. In the 1980s, the US government introduced right-to-know policies to allow the public to learn about toxic discharges from private industries, but the military kept its bases exempt on grounds of national security. If bases had to reveal what substances they were dumping into drains and rivers, it argued, the enemy would be able to use the data to ascertain what operations were taking place at each installation.

Another of its approaches was to claim sovereign immunity; even when breaking environmental laws, it argued, it couldn't be prosecuted since it was a federal entity—and the government can't prosecute itself. Furthermore, it avoids oversight by asserting the right to regulate itself instead of having such standards imposed from outside. The risks of such self-regulation are clear: They prevent public oversight and allow the military to conceal its violations.[17]

This lack of accountability has very real implications for public health. The military has repeatedly been caught covering up the dangers of the substances it uses. In the 1960s, it hid the risks of its defoliants. Likewise, it has ignored or covered up the dangers of perchlorate—a substance found in ammunition and rocket fuel that can interfere with the thyroid gland—and, more recently, carcinogenic per- and polyfluoroalkyl substances (PFAS) found in firefighting foam, which have polluted drinking water sources near US installations around the world.

The military has displayed similar malfeasance by covering up and lying about its environmental compliance. In 1999, thirty-two of the sixty-six former installations the military claimed had been cleaned up were still contaminated with dangerous levels of asbestos, PCBs, and UXO. Instead of conducting full tests on some of the sites, the military had instead merely driven past them or conducted telephone interviews with their caretakers.[18]

Many of the military's lies were related to its dismantling of chemical weapons. In the late 1990s and early 2000s, the military faked safety data and suppressed information that had the potential to alarm the public. Money scandals have also been rife. In 2004, financial inconsistencies in the military's cleanup programs became so widespread that the inspector general for the Department of Defense recommended that the entire US Army undergo ethics training.[19]

At the same time it was attempting to sabotage civilian oversight, the military embarked on a slur campaign against those pushing for more transparency. In 1990, Admiral David E. Jeremiah, the second-highest-ranking military official in the United States, stated that environmental policies often contained hidden agendas that were "antimilitary." Apparently referring to environmental groups, he warned of the risks of a "few local demagogues" damaging military operations. The comments were echoed in 2001, when Vice Admiral James Amerault said environmental laws "provide a powerful weapon for those who oppose the military." Meanwhile, the military allowed false rumors to spread that environmental laws would be so costly that bases would be forced to close and jobs would be lost. Such underhanded tactics have led one top lawyer to label the Pentagon the nation's "premier environmental villain."[20]

To date, the US government has spent $11.5 billion cleaning up the consequences of decades of environmental negligence on military installations in the United States, and in the coming years, it estimates it will need a further $3.4 billion. These numbers are the toxic tip of the iceberg; they only relate to shuttered bases in the United States and do not take into account ongoing contamination from active bases or its past and present operations overseas.[21]

In theory, many of the new environmental policies covered *all* federal installations, including military bases abroad. But in practice, they were not applied due to geographical distances from Washington's bureaucrats and the military's mindset, which compelled it to spurn all civilian encroachment. The military's attitude toward its pollution outside the United States has been even worse than at home. In 1992, following the closure of Subic Bay Naval Base in the Philippines, it repeatedly lied to avoid cleanup, falsely claiming the country had no environmental laws and then saying it was waiting for them to be translated, despite the fact they were already available in English.[22]

In the 1980s the Pentagon maintained approximately 400 bases in 36 countries—including, as of 1985, 105 in Japan, 47 of which were on Okinawa. Operating with zero local oversight, their environmental regulations were controlled solely by the base commander, who wielded carte blanche to guide his installation in any way he chose.

## US FACT-FINDING VISITS FOR OVERSEAS BASES

Between the mid-1980s and early 1990s, US government agencies embarked on at least five trips to overseas military bases to ascertain their environmental conditions. The first of these visits was made by the Office of the Inspector General of the Department of Defense, an agency whose mission is to independently investigate problems in the Pentagon and inform Congress of its findings. Between October 1985 and February 1986, inspectors visited US military facilities in seven foreign countries, presumably including Japan, to check the handling of hazardous materials and waste. Although the full extent of what the team uncovered was not released, snippets made their way into the public realm. One part of the report criticized the bases' handling of hazardous waste; another states, "Policy, guidance, and technical implementation are fragmented, conflicting, and almost nonexistent at installation level." The inspector general of the Department of Defense made two follow-up visits to overseas bases, but again no details were made available to the public.[23]

In 1986, the civilian General Accounting Office (GAO), the body that works for Congress to investigate how the federal government uses taxpayers' dollars, also visited overseas military installations for the first time. It inspected thirteen bases in seven countries, and the findings were published in a September 1986 report titled "Hazardous Waste: Management Problems at DoD's Overseas Installations." Normally such reports are made publicly available, but the Pentagon demanded that this one be classified on the grounds of national security.[24]

Fortunately, parts of the GAO report again made their way into the public sphere. One section states, "We identified actual or potential pollution at eleven of the thirteen bases visited. We concluded that poor hazardous waste management practices and improper disposal could result in water, land, and air pollution." The report concluded that problems with hazardous waste storage "could harm humans and the environment." The names of the countries and bases the GAO visited were not revealed.[25]

Five years after their first report, in 1991, the GAO published a follow-up investigation, a heavily edited version of which was released with the names of the bases and nations removed. According to the report, inspectors visited seven overseas facilities in multiple countries, at least two of them in the Pacific region; one of these seems to have been Kadena Air Base on Okinawa. Within these seven installations, there were more than three hundred contaminated sites, and the report's authors pulled no punches in their criticism of the military.[26]

There was insufficient environmental training throughout the chain of command, and staff did not have enough language skills to communicate how to handle hazardous waste. Five of the seven bases were contaminating the local water system with substances including waste oil and electroplating fluids; solvents from a photography laboratory were being poured on the ground. Inspectors criticized the bases for unauthorized dumping and their practice of mixing hazardous substances with fuel and burning. One host country was apparently so concerned about the storage of hazardous materials that they warned a "major catastrophe" might occur if there was a fire or spill.

The GAO criticized the military's treatment of contaminated soil; after its discovery, it was merely left in place or dug up and reburied on another part of the base. Inspectors censured the selling of hazardous materials to members of the public. For example, at auctions one of the bases grouped hazardous waste and useful chemicals in the same lot to ensure the dangerous substances were sold. Other toxic waste was sold without labels or incorrect markings, so there was no way for the purchaser to know what they were buying.

What must have most worried inspectors were the potential economic costs of the military's overseas environmental damage. As of October 1990, "1,259 claims totaling about $25.8 million" had been received by the military related to environmental problems. Although the military had only paid out about $50,000 in compensation, the GAO estimated that a further three hundred sites within the seven bases had potential for claims, and in the future, the United States might have to compensate for these, too.

The US government and its military were beginning to understand that environmental problems at its overseas bases might begin to hit where it most hurt: the pocketbook. But when it came to Japan, they need not have worried. There they were protected from such financial concerns thanks to a decades-old deal that enshrined the US military's right to operate outside Japanese law.

## THE 1960 JAPAN–US STATUS OF FORCES AGREEMENT (SOFA)

After Japan's surrender in 1945, US occupying forces conducted a purge of militarists and introduced wide, pioneering reforms, including the dismantling of the warmongering *zaibatsu* conglomerates and the redistribution of land. In 1947, these reforms were encapsulated by a new constitution drafted by the occupiers—one of the most progressive ever penned. Granting equal rights to everyone—regardless of sex, race, or social status—it included an unprecedented article—number nine—according to which Japan renounced the right to wage war on other countries. Although written by foreign, occupying forces, the pacifist article was embraced by the Japanese public, tired of a war that had killed 3 million of its nationals; supporters of article nine remain in the majority today.[27]

The 1947 constitution proved to be the high-water mark for US progressivism in Japan. The spread of Communism in China and increasing support for it in Japan convinced the United States to roll back its earlier approach; instead, it took a more reactionary stance. In what is known as the Reverse Course, Japanese war criminals were released from prison and Communist sympathizers fired from their positions in the civil service and private sector. Many Japanese people were dismayed to see some of the militarists who had marched their country to war being released from prison, free to return to high-powered positions. Schools were discouraged from teaching about Japanese military atrocities in China—including the use of chemical and biological weapons—and the media avoided reporting about them; a national narrative took hold depicting Japan as a victim rather than aggressor.

September 8, 1951 was bittersweet for many Japanese people. On one hand, the signing of the Treaty of San Francisco signaled the end of the US occupation of mainland Japan—it took effect the following year—but on the same day, the United States and Japan signed a security treaty that allowed the United States to retain military bases there. The United States envisaged Japan as part of an arc of militarized islands stretching throughout the Pacific to contain the spread of Communism, but many Japanese perceived the military presence—more than twenty-eight hundred facilities in 1952—as a breach of its sovereignty, a feeling heightened by the treaty's severance of Okinawa from Japanese control.[28]

Throughout the 1950s, the CIA pumped millions of dollars into Japan to ensure that the right-wing Liberal Democratic Party (LDP) remained in power; the agency also invested large sums to wage propaganda in support of nuclear power to defuse anti-American sentiments following the *Lucky Dragon* incident. However, by the 1960s, with the security treaty up for renewal, public support for the alliance was only 14 percent; the majority of Japanese people—59 percent—supported a neutral role for Japan, akin to that of Switzerland. In the months leading up to the renewal, 16 million people took to the streets to protest the agreement, anger that later forced President Dwight D. Eisenhower to cancel a visit to the country.[29]

In 1960, Prime Minister Kishi Nobusuke, one of the suspected war criminals released during the Reverse Course, rammed the agreement through Japanese parliament, and the Treaty of Mutual Cooperation and Security between the United States and Japan took effect. Accompanying the agreement was a framework governing the rights and responsibilities of the United States Forces in Japan (USFJ), laid out in a twenty-eight-article document called the Japan–US Status of Forces Agreement (SOFA). For decades, the Japanese people have despised the deal, particularly its treatment of service members who commit crimes against locals, because it grants jurisdiction to military justice proceedings that, when carried out, have been opaque and allowed Americans to avoid punishment under Japanese law.[30]

Just as unfair are the articles related to US military damage to the environment. Article IV relieves the US military of any need to conduct cleanup work on land returned to civilian use or pay for such projects:

> The United States is not obliged, when it returns facilities and areas to Japan on the expiration of this Agreement or at an earlier date, to restore the facilities and areas to the condition in which they were at the time they became available to the United States armed forces, or to compensate Japan in lieu of such restoration.

Moreover, Article XVIII exempts active US service members who damage Japanese property from any claims of compensation:

> Each Party waives all its claims against the other Party for damage to any property owned by it and used by its land, sea, or air defense services, if such damage a) was caused by a member or an employee of the defense services of the other Party in the performance of his official duties.

In the United States, legal action against service members who dumped hazardous waste had been one of the ways the courts ensured compliance; however, SOFA effectively ruled out similar prosecutions in Japan.

Both clauses were inked in 1960, at a time when there was international lack of understanding of human impact on the environment. In the ensuing decades, research on *industrial* contamination, for example, from methyl mercury poisoning in Minamata, Japan, and *military* contamination, for instance, the Camp Lejeune water scandal, has revealed the damage we cause to the environment and ourselves. But despite this decades of accumulated knowledge, Articles IV and XVIII of SOFA have never been revised. The two clauses exempt the US military in Japan from all environmental responsibility. In effect, SOFA offers the US military the freedom to pollute its bases and, when contamination makes them too toxic for continued use, return them to Japan to clean up.

One more article of SOFA strongly impacts how the US military deals with the environment in Japan. Article XXV established the US–Japan Joint Committee to manage the day-to-day issues that arise from the presence of the US military there: "The Joint Committee shall determine its own procedures, and arrange for such auxiliary organs and administrative services as may be required." Self-regulating, the committee acts in total secrecy with zero public oversight. It meets every other Thursday at 11:00 a.m., alternating venues between the military-owned New Sanno Hotel in Tokyo and a location chosen by Japan's Ministry of Foreign Affairs (MOFA). On the US side, there are seven members—six of whom, including the main representative, are appointed by the military; the seventh is the US Embassy's political-military representative. Such a makeup gives the military overwhelming power over their State Department counterparts and frees it from the civilian interference it so despises. On the other hand, the Japanese side is headed by the MOFA North American top and five representatives from the ministries, including only one from the Ministry of Defense. In diplomatic terms, the majority US military presence is an abhorrence. Usually the civilian State Department handles international diplomacy, but the committee's military dominance makes it resemble an ongoing occupation. There is no obligation for the committee to make its decisions public, and only rarely are they revealed.

Such secrecy has harmed people living near bases—especially on Okinawa—in very real ways. In November 1973, there was an agreement titled "Memorandum to the Joint Committee on Cooperation Concerning Environmental Matters." It stated that "as a principle, pollution problems be solved through local initiative." To achieve this, prefectural or municipal authorities would be allowed—via the Japanese Defense Bureau—to ask the base commander to investigate a suspected leak of oil or chemical substance, and the results would be given to the authorities as quickly as possible. The agreement also allowed local authorities to apply to the base commander to inspect the suspected source of contamination firsthand and conduct their own samples.

Granting local authorities the right to inspect the bases in their communities, the agreement gave them concrete powers to protect their environment and health of residents. The memorandum applied to all of Japan, but it was particularly important for Okinawa—because of not only its large number of bases, but also the timing; coming a year after the island's return to Japan, it was a major step toward making the military more accountable in the newly reestablished prefecture.

But the 1973 agreement was kept hidden for thirty years. It was not until 2003, that the Okinawa Prefectural Government learned of its existence. Asked why it hadn't been revealed earlier, Japan's foreign minister explained that withholding the information would not create any problems. This couldn't be further from the truth. In the short term, failure to disclose to local authorities their right to inspect bases prevented investigations on Okinawa into the usage of defoliants against Iejima's farmers in September 1973. Then it hobbled their ability to investigate the August 1975 spill of carcinogenic detergents at Makiminato Service Area and the spate of fuel leaks from US bases that contaminated the sea, rivers, and farmers' fields in the mid-1970s. In the longer term, hiding the agreement allowed the US military on Okinawa to continue unsafe practices without fear of being held legally accountable.[31]

For many years, Okinawans had campaigned to become a prefecture again, with the expectation they would gain the protection of Japanese laws. But the hiding of the 1973 agreement was another sign that the bases would stay exempt from legal responsibility; the island remained a place where things could be done that could not be done elsewhere—and now the Japanese government had become complicit in that US military exploitation. The hiding of the memorandum set a post-reversion precedent for the suppression of Okinawans' right to know by both the Japanese government and the US military. It also signaled that the Japanese government cared little for the safety of residents living near mainland bases who had suffered from environmental accidents throughout the Cold War. In the coming years, the situation became even grimmer, as more environmental policies were introduced that served not to protect the rights of residents but to guarantee the military's ability to continue operating without oversight.

## A GAP-RIDDEN PATCHWORK
## OF ENVIRONMENTAL POLICIES

Today, the environmental policies covering US military bases in Japan consist of a patchwork of Department of Defense directives and agreements worked out behind closed doors of US–Japan Joint Committee meetings. Realizing that overseas base pollution might lead to economic and political problems, in 1991, the Department of Defense created its first attempt at uniform environmental standards on foreign bases, Directive 6050.16—Department of Defense Policy for Establishing and Implementing Environmental Standards at Overseas Installations; it has received intermittent cosmetic updates. The policies claimed to "establish and maintain a baseline guidance document for the protection of the environment at DoD installations," for example, how military bases handle and dispose of hazardous substances; they also laid out how the host nation ought to be notified in case of a spill.

In Japan, the guidance document is known as the Japan Environmental Governing Standards (JEGS). The first of these was drawn up in 1995, and the most recent is dated April 2018; their contents are guided sub rosa by the Environmental Subcommittee of the Joint Committee. As such, how JEGS are determined is not open to public scrutiny, and, unsurprisingly, given the US military dominance on the committee, the guidelines are heavily weighted in the Pentagon's favor.

The first failure of JEGS is they lack teeth, specifically stating that they do not create any enforceable rights against service members or the US government. Without fear of punishment, there is no motivation for bases to follow these guidelines, so service members continue to hide spills, delay reporting them, play down their severity, or conceal the dangers of the substances involved. Another problem is that the military has been able to keep certain hazardous substances from the lists of those needing reporting, for example, PFAS and depleted uranium. Moreover, the policies exempt military aircraft and vessels from environmental compliance.[32]

Worsening the failure of JEGS is the free pass they give the military for its decades of past contamination. The policies do not provide for contamination checks, nor do they cover its cleanup. As a result, unlike in the United States, service members and neighboring residents have no idea how polluted installations are. On mainland Japan, this is particularly troubling because many of today's US facilities used to possess large stockpiles of chemical weapons when the bases belonged to the Japanese military.

Accompanying JEGS, other policies have been worked out in secret at US–Japan Joint Committee meetings. In September 2000, the governments of Japan and the United States issued their "Joint Statement of Environmental Principles" (JSEP), stating that the US government would "remedy any contamination caused by USFJ that poses a known, imminent, and substantial threat to human health."[33]

Stemming from 1990s Pentagon guidelines, the use of the words *known, imminent, and substantial* is crucial because it effectively provides an incentive for the bases *not* to conduct tests for contamination. If they do check and pollution is subsequently discovered—i.e., it becomes "known"—they have to remediate the problem. This desire for contamination not to become "known" also extends to US authorities' failure to inform the media of environmental accidents and aggressive clamp downs on journalists' reporting on such matters.

Equally troubling is the inclusion of the words *imminent* and *substantial*—both of which are left solely to the base commander to define. The danger of many environmental toxins is their ability to accumulate in the human body over time—a risk excluded by the word *imminent*. Likewise, the definition of the term *substantial* lacks any independent scientific oversight to determine exactly how concentrated levels of contamination must reach to fall into this category. Compliance with JSEP—like JEGS—is entirely at the discretion of the base commander, so those on the Japanese side, including prefectural and municipal authorities, have no say in its execution.

In December 2013, likely in response to public anger following the discovery of the 108 contaminated barrels in Okinawa City, the Japanese government announced—to great fanfare—its upcoming Framework Regarding Environmental Stewardship; however, any hopes of significant reforms were soon dispelled by the inclusion of the spineless statement, "The Government of Japan affirmed its responsibility for environmental restoration."[34]

When the new agreement—Cooperation Concerning Environmental Matters—took effect in September 2015, Tokyo and Washington claimed it was historic, but it changed nothing. It merely gave Japanese authorities the right to *request* inspections following a toxic spill or upcoming return of land. Nonetheless, such permission remains at the discretion of the US military, which can deny access on the broad grounds that it would "interrupt military operations, compromise force protection, or hinder the management of the facilities and areas."[35]

## OTHER COUNTRIES ASSERT THE NEED FOR US MILITARY ACCOUNTABILITY

Compared to the submissive attitude of the Japanese government, other countries have been more assertive in holding US military bases within their borders responsible for environmental damage. In Germany, for example, the nation's SOFA has an environmental clause that makes the German government responsible for only 25 percent of cleanup costs. As far as noise pollution is concerned, bases have been pressured into setting up a complaint hotline and holding regular noise reduction meetings with nearby residents. Also, in 2016, the US Army agreed to investigate PFAS contamination near its base in Ansbach. Following the discovery of high levels of the harmful substances, the military acted to remediate the contamination—in stark contrast to its response on Okinawa, as the following chapter shows.[36]

Likewise, in South Korea, there have been pushes to make the US military more accountable. Following the return of twenty-three bases without cleanup in 2007, the Korean government pushed for a Joint Environmental Assessment Procedure (JEAP), which was established in 2009. According to its provisions, the US government is required to reimburse cleanup costs associated with contaminated former military land. Also, after an incident in February 2000, when a US soldier dumped formaldehyde into Seoul's Han River, the US Army promised to spend $100 million to improve its environmental activities.[37]

Within the US government, there has been some recognition of the need to remediate overseas environmental damage—if only for political reasons. In Vietnam, the United States remediated dioxin contamination at Da Nang Airport and moved on to clean up the former Bien Hoa Air Base, near Ho Chi Minh City. Undoubtedly, these cleanups are too little too late, but their very existence illustrates how Washington is finally acknowledging the environmental damage of its military's actions, even going as far as to help a former enemy. When it comes to its self-professed strongest ally, Japan, however, no such assistance has been forthcoming. Perhaps this is most apparent at the facility that ranked alongside Da Nang as one of the busiest during the Vietnam War: Kadena Air Base, similarly contaminated with military herbicides and dioxin.

At Da Nang Airport, workers wore full hazmat suits, and the technology to remediate the contamination was state-of-the-art, baking the soil at high temperatures to destroy the dioxin. Signs posted in the vicinity clearly stated the dangers of entering the area. In contrast, the cleanup at the Okinawa soccer pitch was shockingly slipshod. Throughout the years-long project, cars and pedestrians freely passed within meters of the dusty pit, and workers often labored without hazmat equipment. Following a typhoon in July 2015, water from the flooded site was pumped into local culverts without any safety checks. At a local level, the scene highlighted potentially life-threatening ignorance of the dangers of military contamination. At a deeper level, it embodied the Japanese government's failure to hold the United States accountable for its environmental damage. Tokyo has the option of pushing the US military for more responsibility, but it chooses not to. This combination of Japanese complacency, military arrogance, and Joint Committee opacity came together in a perfect storm of contamination, this time not caused by the military, but by a private Japanese company.

## ATSUGI AIR BASE AND THE SHINKAMPO INCINERATION COMPLEX

Six thousand American service members and their families, as well as Japanese civilians, work at Naval Air Facility Atsugi (NAFA), located forty kilometers southwest of Tokyo. Between 1985 and 2001, the privately run Shinkampo Incineration Complex operated 250 meters outside the base, burning about ninety tons of medical and industrial waste per day. Situated in a deep valley, its three chimneys pumped smoke directly level with the installation's housing quarters and child care center.[38]

Laurie Paganelli, a NAFA resident, described the smoke as follows: "It smelled, burned your eyes, and sometimes added a greenish glow to the air around us." Others complained of headaches, nausea, and breathing problems. The brunt of this exposure was borne by dependents—whereas sailors were at sea for long periods, their families remained on the base, breathing the fumes for their entire three-year deployments.[39]

Contained in the smoke were PCBs, pesticides, heavy metals, and dioxin; the Department of the Navy described it as a "witch's brew of toxic chemicals." Within the installation, dioxin readings for soil contained the highest levels ever recorded in Japan: fifty-three picograms per cubic meter.[40]

In 1988, base residents began to complain about the smoke, prompting the US Navy to introduce air quality surveys. Four years later, in 1992, the United States raised the matter at the US–Japan Joint Committee, but due to the opaque nature of the proceedings, no warnings were made to base residents.

In 1995, the navy detected twelve emissions exceeding EPA levels at the installation, and the following year, a special form was added to residents' medical records notifying doctors that they had been exposed to carcinogenic and other hazardous substances. However, the navy continued to downplay the dangers; in 1997, the commander told residents wishing to relocate that there was not "anything in writing that says Jinkampo [sic] is a serious health risk!" Later that same year, the installation proffered only rudimentary advice, for instance, keeping children indoors when smoke blew toward the base.[41]

In 1998, the navy introduced a waiver—reportedly the only one in the world—that residents had to sign upon arrival at NAFA. The form stated, "The Jinkanpo [sic] incinerator produces a plume of air pollutants, including dioxins, that frequently blows directly over NAF Atsugi, including the housing and child care areas." The navy did not include any explanation of the illnesses potentially caused by exposure, nor did it give service members and their families the option to live off of the base.

Japanese authorities were equally negligent in ensuring the safety of those at NAFA. They reacted with skepticism to the results of the navy's environmental surveys and gave only recommendations to the incinerator's owner, for example, installing filters on the chimneys, but there was no order for the complex to cease operating. The owner of the incinerators had deep ties to the Yakuza and frequently thwarted attempts to monitor his plant.[42]

In 2001, after thirteen years of military inaction, the US Department of Justice sued the owner of the incinerator, charging that the complex endangered the health and safety of those on the base, and "interfered with US government rights of property and possession." As a result, the Japanese government paid the incinerator's owners $42 million to close the site.

More than twenty-five thousand service members and their dependents were exposed to contamination at NAFA, but today the VA states, "Currently there is no definitive scientific evidence to show that living at NAF Atsugi while the incinerator operated caused additional risk for disease." Service members cannot receive automatic help, and claims are decided on an individual basis; their families are not eligible for compensation.[43]

Former residents are certain of the health impact of breathing Shinkampo's witch's brew. In 2009, Paganelli, who described the smoke's "greenish glow," testified in front of Congress that there had been at least 61 cancer cases in a group of 750, one of which was her own son. "We trusted the navy to provide a safe environment for our family members, but they failed to do so, knowingly housing our families in a toxic waste zone," she said.[44]

In 2012, the GAO criticized the Department of Defense's handling of the Atsugi case, including the failure of existing policies to protect service members' dependents on overseas bases. As a result of the GAO report, the Department of Defense said it would include dependents to improve the scope of its surveys—but as shown by the discovery of 108 contaminated barrels near Kadena Air Base, such measures are still sorely lacking.[45]

## POISONED CHALICES

In the past four decades, the Department of Defense has gradually scaled back its presence in Japan. In 1972, at the time of Okinawa's reversion, there were 87 US military installations in the prefecture; today there are 31. During the same period on mainland Japan, the number of bases decreased from 103 to 47.

Returns of military land ought to be moments to celebrate for local communities who have waited long to put it back to use, but properties have been so badly polluted that cleanup has taken years or decades to complete. Until the 1990s, Japanese authorities conducted few environmental checks on land returned from military control; as environmental awareness grew and more tests were conducted, serious contamination was detected on mainland Japan.[46]

As worrying as these mainland discoveries have been, on Okinawa the contamination has been even more severe and widespread; as of 2018, Japan had paid 12.9 billion yen ($117.3 million) for the cleanup of contamination and other remedial work on former base land. For many people, the first eye-opener followed the closure of Onna Communications Site in November 1995. In the past, the base had caused several spills of sewage and detergents, but since there had been no major industrial operations at the site, fears of contamination were low. When Japanese authorities conducted environmental tests, they discovered cadmium, lead, arsenic, and, from sewage disposal tanks, 104 tons of PCB-contaminated sludge. In 2002, 218 tons of similar sludge was discovered on a nearby Japan Self-Defense Forces subbase that had been taken over from the US military in 1973. The two discoveries filled 1,794 barrels.[47]

The contamination set alarm bells ringing in Washington. In a 1998 report titled "Issues Involved in Reducing the Impact of the US Military Presence on Okinawa," the GAO realized that the trouble at Onna was a taste of things to come. When military land is returned in the future, "If a survey is conducted and contamination is found, a decision would be needed as to whether the United States or Japan would pay cleanup costs," the report stated.[48]

Raising as it does the potential that the United States might be willing to bear responsibility for remediation costs, the statement ought to have been seized upon by Tokyo to push for US military accountability. But either due to the Japanese government's ignorance of pollution risks and/or its fear of upsetting Washington, Tokyo elected to remain silent and allow the existing, unfair system to prevail.

It took seventeen years for the contamination at Onna to be remediated—and the final solution angered many people. Starting in November 2013, the barrels were transported for disposal to Fukushima Prefecture, where, two years earlier, nuclear meltdowns had spewed high levels of radiation. Now US military contamination was being added to this mix.[49]

At about the same time as the closure of Onna Communications Site, Washington and Tokyo announced large land returns on Okinawa. Their decision had not come willingly. On September 4, 1995, a twelve-year-old girl had been abducted and raped by three US service members from Camp Hansen. In the weeks following the attack, the commander of the United States Pacific Command, Admiral Richard Macke, publicly mused that hiring a prostitute would have been cheaper than renting the car the men used to kidnap the girl; he was fired. For the first time since the late 1960s, fury on Okinawa grew so strong that Washington and Tokyo feared it might jeopardize the security of the bases there.[50]

Forced into concessions, the two governments established the Special Action Committee on Okinawa (SACO), and, in December 1996, it announced eleven parcels of military land—totaling fifty square kilometers—would be returned. Among these would be parts of the Northern Training Area (NTA), Naha Military Port, Makiminato Service Area, and, ultimately, Marine Corps Air Station (MCAS) Futenma. More transparency regarding Joint Committee decisions was also promised.[51]

Japan and the United States heralded the returns as major steps toward lessening Okinawa's military burden, and the announcement succeeded in temporarily subduing anger on the island. But slowly the reality of the so-called reductions became clear: The returns were tied to conditions that made them military upgrades in disguise. For example, releasing half of the NTA would require building new pads for USMC Osprey aircraft, and the closure of MCAS Futenma would only occur in exchange for the construction of a new base, burying the pristine coral reefs of Oura Bay.

In recent years, Okinawa has become one of the most popular tourist destinations on the planet; in 2017, the number of visitors—9.4 million—almost matched that of Hawaii. Okinawa is no longer dependent on the US military to keep its finances afloat; today the bases contribute only 5 percent to the prefecture's economy, compared to 15 percent in 1972. Such a statistic is often lost on Japanese and American military proponents whose mindset dates back to the pre-reversion era, when the island still used dollars and tens of thousands of Okinawans were employed by the bases. Once a boon to the island's economy, the military presence has now become the primary impediment to development.

When property returns have eventually materialized, extensive contamination has been discovered. Some of the most serious problems have occurred in Chatan, where thousands of buried bullets, parts of a tank, and elevated levels of benzene, arsenic, hexavalent chrome, and lead were found on former parts of Camp Kuwae. Located close to a popular tourist area, past redevelopment of military land there in the 1990s resulted in economic improvements 108 times over its former usage. But similar profits in recent years have been stymied by contamination; at the time of this writing, fifteen years after the land's return, it still has not been fully redeveloped.[52]

March 2015 witnessed the closure of the military's Nishi Futenma housing area, located atop a hill with picturesque views of the East China Sea.

Once again, however, the subsequent discovery of serious contamination dampened the celebratory mood, as high levels of lead, arsenic, and oil were found throughout the property. Also discovered at Nishi Futenma were four sites contaminated with the solvent dichloromethane, which can cause chemical burns and has been linked to cancer. That such serious contamination existed at a former housing area with little or no industrial use underlines how US military operations pollute all kinds of land. The Japanese government's failure to be more assertive in protecting the rights of Okinawan residents flies in the face of the United States, where the government has frequently confronted its own military concerning environmental infractions. In the case of Nishi Futenma, it also shows the disdain with which the military treats the health of service members and their families. Even though former US residents may have been sickened by the contamination beneath their homes, no notifications have been made to them, and unlike in the United States, where such information is readily available on EPA websites, no such resources exist in Japan.[53]

Even the CIA has been advising the US military to be more careful about its treatment of the environment on Okinawa, albeit for its own ulterior motives. In December 2017, via the Freedom of Information Act (FOIA), I received a sixty-page manual from the CIA division the Open Source Center, called "Understanding Base Politics in Okinawa." Dated January 2012, the manual aimed to advise US policy makers on how to manipulate Okinawan public opinion about the military presence in their prefecture.[54]

According to the CIA, "Okinawan support for environmental preservation presents challenges to alliance managers." It warned policy makers that "Okinawa may pressure Tokyo to expand environmental guarantees for base land—including revision of SOFA provisions on environmental remediation." To improve Okinawans' support for bases in the prefecture, the CIA recommended that the military respond to environmental incidents quickly and transparently.

The Okinawa manual was part of a series called "Master Narratives," in which other reports focused on Syria and Afghanistan. The fact the CIA saw Okinawa among such trouble spots is very revealing. At the time it was written in 2012, the LDP, long-backed by the CIA and other US government agencies, had been booted out of power for only the second time since 1955. The new party in charge, the centrist Democratic Party of Japan (DPJ), had promised to be more assertive with the US military, for example, canceling plans for the new USMC facility at Oura Bay; the creation of this manual suggests the CIA saw the DPJ as a very real threat to US military dominion over Okinawa. DPJ rule only lasted three years, however, and, in 2012, the LDP returned, headed by Abe Shinzo, who, like his grandfather, Kishi Nobusuke, displayed total subservience to US military demands and equal disdain for democratic processes. At the urging of the US government, Abe bulldozed through parliament a state secrets act that threatened whistleblowers and journalists with long imprisonment, and a collective self-defense act that allowed Japanese troops to fight overseas with US troops and multitrillion yen deals to buy US weaponry.

Unsurprisingly, LDP obsequiousness extended to measures countering US military contamination in Japan, which, already almost nonexistent, became even more lax. Since 1978, the Ministry of the Environment had conducted rudimentary environmental monitoring each year on bases throughout Japan, checking sewage plants and boilers for harmful emissions; however, from 2014 onward, these checks were halted due to a change in policies by the Joint Committee. When journalists pressed the government for the reason, the Ministry of the Environment said it could not answer because it was a result of a decision between the US and Japanese governments.[55]

Given the twin obstacles of physical barbed wire fences and opaque policies maintained by Washington and Tokyo, those of us wishing to know about the extent of military contamination in Japan need to rely on more creative approaches. Fortunately, two options are available: FOIA and whistleblowers brave enough to risk the wrath of the military for whom they work.

# 6

# OKINAWA

Paradise Lost

**D**espite Japanese government promises prior to reversion that Okinawa's military burden would be reduced to the same level as on the mainland, today 70 percent of US bases in Japan are crammed onto the small island, thirty-one facilities occupying some of its best farmland and prime real estate. Not only has the US physical occupation of Okinawa persisted, but also its colonial mentality. In recent years, high-level US officials have maligned the island's residents in a series of well-publicized cases. In 2010, senior diplomat Kevin Maher allegedly accused Okinawans of being masters of manipulation and "too lazy to grow goya"; then, in 2015, Robert Eldridge, deputy assistant chief of staff of government and external affairs, appeared on a Japanese neo-nationalist YouTube channel, where he accused antibase demonstrators of engaging in hate speech against Americans—several weeks later, he was fired after a leak of military surveillance footage to a presenter at the same network. In 2018, the head of the United States Marine Corps (USMC), General Robert B. Neller, infuriated Okinawans when he asserted that no one had lived on the land where Marine Corps Air Station (MCAS) Futenma was built, whitewashing the town's history of internments and forced relocation by the United States.[1]

Today, the military presence on Okinawa is dominated by the USMC, whose eleven bases occupy diverse environments. MCAS Futenma sits in the urban midst of Ginowan City surrounded by more than 120 public facilities, such as schools and hospitals. Camp Hansen sprawls across dense jungle through which three main rivers run and two dams store water for local communities. The adjacent Camp Schwab occupies a large stretch of coastline, including Oura Bay, one of the most biodiverse places in Japan. Other USMC bases include Iejima Auxiliary Airfield, the Northern Training Area (NTA), and Makiminato Service Area, transferred from the US Army in 1978, and renamed Camp Kinser, after a marine sergeant who had flung himself on a Japanese grenade during the Battle of Okinawa.

In and around these bases, there are 260 rare, threatened, or endangered animal and plant species, notably the dugong, a cousin of the manatee, and the flightless Okinawa rail. Military public relations campaigns project the image of the USMC as a careful custodian of the island's environment; several times a year, service members collect garbage from beaches and base fences; in 2015, the USMC on Okinawa won a top award for environmental quality from the Secretary of Defense. As well as targeting the Japanese public, such campaigns aim to convince Americans that the bases on which they work and live are clean. Military members and their families are aware of the contamination of Department of Defense facilities in the United States, and so the Pentagon is determined that no suspicions arise as to the safety of those on Okinawa. Japanese requests for environmental tests can be ignored, the military knows, but such demands from within its own ranks might need to be taken more seriously, and if contamination were to be found, it would need to be remediated according to guidelines governing "known" pollution. Because of this, the USMC, like other military branches on Okinawa, works hard to conceal information related to its contamination; however, the US Freedom of Information Act (FOIA) offers one way to cleave through such secrecy.[2]

## CHEMICAL CONTAMINATION AT CAMP KINSER

Throughout the 1960s and 1970s, evidence suggested serious environmental problems at Makiminato Service Area, the storage site for materiel returned from the Vietnam War. Large chemical spills occurred in 1975, particularly the one that exposed base workers to carcinogenic industrial detergents, dismissed by the US Consulate as a "flap." Further instances in the following years suggested environmental safety at the facility was lax, but due to the Status of Forces Agreement (SOFA) blocking on-base access, there was no concrete proof.[3]

In 2014, I received a tip-off from a military source that the USMC possessed a package of documents related to contamination at the installation, now called Camp Kinser. I filed a FOIA request, but for the next sixteen months, the USMC blocked the release; it delayed processing my request, then refused to hand over the report on the grounds of protecting against "public confusion." Finally, in a last-ditch attempt to keep the documents under wraps, it suddenly claimed it didn't have them in the first place.

Realizing the only way to obtain the reports was to publicly shame the USMC into releasing them, I wrote a story about their shenanigans that ran in the English newspaper the *Japan Times*. Less than a week later, the USMC coughed up the reports in full.[4]

Produced between 1977 and 1993, the eighty-two pages were the first internal documents related to US military contamination in Japan ever made public—and, upon reading them, it soon became clear why the USMC had tried so hard to keep them secret. According to the reports, throughout the 1960s and 1970s, the shoreline of the base consisted of a storage yard measuring forty-six thousand square meters, containing supplies returned from Vietnam described as "insecticides, rodenticides, herbicides, inorganic and organic acids, alkalis, inorganic salts, organic solvents, and vapor degreasers." The military had attempted to auction off some of these stocks to Okinawan civilians, but when the winning bidders turned up, the containers were in such poor condition they refused to take them. Okinawa government officials visiting the site in the mid-1970s noted hundreds of barrels, boxes, and plastic bags leaking chemicals in the area, but the military explained they held malathion, an insecticide that, while toxic, was not as dangerous as their real contents.

In 1974 and 1975, large "fish kills" on the nearby coast prompted the US Army to conduct surveys of the sea and soil. The results revealed high levels of polychlorinated biphenyls (PCBs) and a "high concentration" of dioxin, later specified as an "Agent Orange component." Tests in 1978 also revealed elevated levels of the carcinogenic heavy metals, lead and cadmium. To clean up the site, the military buried large quantities of chemicals on the base, including sludge from treated cyanide compounds and 12.5 tons of ferric chloride, a highly corrosive industrial chemical. Pesticides were also taken north to Camp Hansen and buried there.

In the mid-1980s, toxins again seeped from the base, resulting in the further death of marine life. At this time, military personnel criticized their predecessors for their failure to conduct follow-up checks or record whether the contaminated soil had been removed from the installation. An internal report from 1984 raises concerns that construction crews might have been exposed to PCBs, but the military never made any effort to inform them. That same year, high levels of heavy metals were detected on the shore more than two kilometers from the former storage site, and, in 1990, suspected toxic "hot spots" still existed within Camp Kinser, forcing the USMC to abandon plans to build a pleasure beach there.

At the time, the US Navy estimated that a full survey of the former storage site would cost more than $500,000, but an actual cleanup would cost much more. According to the reports, such funding would be difficult to obtain because Pentagon remediation budgets were reserved for projects in the United States.

Included in the FOIA package were charts pinpointing the contaminated area. Crosschecked against current maps of Camp Kinser, today a bowling alley, medical center, and baseball field sit on the site. Additionally, civilian reclamation work has since filled in the sea near the former storage yard, and it now holds shops, offices, and factories.

According to the 1996 Special Action Committee on Okinawa (SACO) and follow-up agreements, Tokyo and Washington intend to return the whole of Camp Kinser starting in 2024 or later, but FOIA requests for recent contamination surveys come back with responses that no records exist, suggesting one of two things: No such environmental checks have been conducted or the military is lying to me.

With no access to Camp Kinser, Japanese scientists have had to rely on checks of wildlife caught nearby to determine levels of contamination. Mongooses caught in the vicinity show high levels of PCBs, and habu snakes contain elevated concentrations of PCBs and DDT. In 2016, sediment in a river flowing near the former storage area was found to be contaminated with the same pesticides described in the 1970s report and high levels of lead; in 2017, lead was also detected in shellfish and seaweed near the base. It seems certain that Camp Kinser is still contaminated with the chemicals brought back from the Vietnam War more than forty years ago.[5]

## USMC ENVIRONMENTAL ACCIDENTS

The SACO final report promised more transparency regarding Joint Committee agreements and, by definition, the environmental decisions made within them; however, despite such assurances, internal documents released via FOIA show that the USMC still believes it can operate on Okinawa without openness. USMC handbooks from 2013 and 2015, reveal that marines are under orders not to inform Japanese officials of "politically sensitive incidents." Since the decision whether to classify an incident as "politically sensitive" is entirely left to the USMC, this gives it the ability to hide any accidents it chooses. Other FOIA-obtained reports show the USMC repeatedly violates environmental guidelines on its Okinawa bases.[6] Such self-regulation leads to sloppy operating procedures. According to reports obtained via FOIA, between 2002 and 2016, there were 270 environmental accidents at MCAS Futenma, Camp Hansen, and Camp Schwab, but only six were reported to Japan.[7]

Even when the USMC decides to report incidents to Japanese authorities, it misleads them. In June 2016, an accident on MCAS Futenma spilled almost seven thousand liters of aviation fuel. The internal report stated the accident was due to human error, but United States Forces in Japan (USFJ) told Japanese authorities it occurred because of a "valve misalignment." Moreover, although USFJ claimed the spill had been dealt with "immediately," the in-house documents reveal it wasn't fully under control until the following day. The scale of the accident was so large it required the disposal of eleven barrels of contaminated earth and three thousand liters of contaminated water.

In 2016, I was contacted by a military whistleblower who had been working on USFJ installations for many years and wanted to share information about safety breaches on the USMC's Okinawa bases. He explained how the June 2016 incident had been caused by marines overriding a safety solenoid using the kind of plastic tie used to seal food bags. "Such accidents are typical of the US Marines. To put it bluntly, their work is lazy, and they act stupidly," he said. The expert provided a video of the spill, showing large volumes of fuel gushing out of a vent in the side of the grass-covered storage tank, pooling on the ground and running into a storm drain. In March 2009, the same fuel tank had suffered a large leak due to similar ineptitude.[8]

Reports also reveal the careless storage of chemicals on Camp Hansen. One incident in December 2011, involved calcium hypochlorite bleach powder sloppily stored in a shipping container. Some of the chemical began to react with the air, injuring a marine who opened the container's door. The incident was kept secret for a month, and after superior officers had finally been notified, the base declared an emergency, calling in a hazmat team from the Japanese fire department to clean up the spill site. In the United States, such negligence led to the prosecution of service members; however, guidelines in Japan give them immunity from punishment under the law.

## TEAR GAS, RED EARTH, AND FOREST FIRES

As well as environmental spills, USMC bases on Okinawa have caused other problems for neighboring communities, including tear gas leaks, red earth runoff, and forest fires. Days after the start of the Vietnam War, Okinawan children were exposed to tear gas at Ginoza Middle School. In the following years, other incidents involving the substance took place, some involving drunk troops. In one such case in November 1968, approximately one hundred households in Kin Town had to flee when a US service member tossed a tear gas canister into a civilian area. Moreover, following reversion, in March 1980, patients and staff from the Ryukyu Psychiatric Hospital were sickened when tear gas leaked from Camp Hansen. And in June 1988, two civilians in a car driving along a prefectural road near the NTA fell ill when they were similarly exposed.[9]

One of the most visible signs of environmental damage from USMC bases has been the runoff of soil from live-fire sites. This has been especially harmful at Camp Hansen, where decades of heavy artillery training denuded hillsides and made them vulnerable to erosion. This runoff, contaminated with lead and toxic propellants, damaged local rivers and the sea. According to internal USMC reports, efforts to reduce erosion by planting vegetation have been hampered by the ongoing presence of unexploded ordnance (UXO).[10]

In addition, between 1972 and the end of 2016, operations at Camp Hansen and Camp Schwab caused almost six hundred forest fires. Many of these fires have been caused by the USMC breaking its own rules. In 2000, the USMC on Okinawa created guidelines that prescribed the conditions in which live-fire drills could be conducted and which types of ammunition could be used; in conditions categorized as "dry," no tracer rounds could be fired, and in "very dry" conditions, training would not be conducted at all. However, according to research by the nonprofit Peace Depot between 2002 and 2007, more than half of the fires—55 percent—occurred after the USMC had ignored its own guidelines. Many of these fires had been due to the use of tracer rounds.[11]

Ammunition was the root of problems at another USMC base, Camp Courtney. For thirty-seven years until its closure in 1999, part of the installation near the shoreline contained a skeet shooting range. When Okinawa prefectural officials learned of the use, they grew concerned that nearby areas growing hijiki seaweed may have been contaminated with lead, so they demanded that the USMC investigate. The military admitted that approximately forty-nine tons of lead was scattered around the site, with the areas in which hijiki grew having elevated concentrations. In 2001, the Japanese government conducted tests in the area and claimed the contamination posed no risk to human health; however, when the prefecture requested to conduct its own independent tests, it was refused. It was not until June 2011, that the US–Japan Joint Committee approved the request, by which time levels were declared safe.[12]

The weaknesses of current environmental guidelines arose again at Camp Hansen in August 2013, when an HH-60 Pave Hawk helicopter crashed near Okawa Dam, a source of drinking water for nearby Ginoza Village. Concerned it may have been contaminated, local officials requested to visit the site but were denied. Unsure whether the dam had been polluted, the village decided to halt drawing water from it. Subsequent tests by the US military of the soil at the crash site revealed levels of arsenic twenty-one times the safe level, as well as high concentrations of cadmium, fluorine, and lead.

A similar problem occurred in October 2017, when a USMC CH-53 Super Stallion helicopter crash-landed and caught fire in a farmer's field near the NTA. Even though the accident happened outside the base, the military cordoned off the site and blocked access by Japanese officials. Before local authorities could conduct environmental tests, the military removed the wreckage and dug up the soil. Belated checks by the Japanese government revealed contamination from benzene and radioactive strontium-90, used in helicopters' rotor devices. As the residents of the Marshall Islands know too well, following exposure, the human body incorporates the substance into the teeth and bones, where it can lead to cancers. The military claimed the strontium had been removed within three days of the crash, but a whistleblower provided me with photographs showing it had remained at the scene for at least a week.[13]

## IN THE CROSSHAIRS OF THE USMC

In 2016, I discovered my investigations into environmental contamination on Okinawa had made me a target of the USMC. With a FOIA request, I obtained a file that showed I was being monitored by the USMC Criminal Investigation Division. The report featured my photograph, biography, and an account of a lecture I'd given outside Camp Schwab about military contamination. Internal USFJ e-mails—again obtained via FOIA—also labeled me "adversarial" and said my "tone of reporting is hostile." At about the same time, I realized the United States Air Force (USAF) was blocking my home ISP from accessing its homepages—an apparent attempt to hobble my ability to file FOIA requests via its online system. Other Okinawan journalists had experienced the same denial.

When these incidents became public in October 2016, the US Freedom of the Press Foundation, cofounded by Daniel Ellsberg and headed by CIA whistleblower Edward Snowden, condemned the US military's monitoring of me; the French nongovernmental organization Reporters Without Borders wrote a letter of protest that was widely circulated. My hometown member of parliament even contacted the British foreign minister to urge the UK government to file a complaint with Washington.[14]

This was not the first time my journalism had brought me into conflict with the military, specifically the USMC. In the spring of 2016, my reporting had made national headlines when a FOIA request had netted me the scripts and slides of the orientation lectures the USMC gives new arrivals to the island. The lectures contained lies and comments disparaging Okinawans, accusing them of having "double standards" and stating their complaints against hosting the burden of US military bases in Japan were "more emotional than logical." They also told new marines that for Okinawans, "It pays to complain. Anywhere offense can be taken it will be used."[15] Likewise, the orientation lectures contained misleading statements about the military's impact on the island's environment: "Some believe that the SOFA does not ensure the US military is abiding to modern environmental standards, and as a result, there is a perception and eagerness to find evidence that our activities are damaging the natural environment."[16]

The cover-up of USMC environmental damage on Okinawa is an appalling disservice to its own troops. Marines—like Okinawans—have a right to know whether their homes and workplaces are poisoning them and their families, but the orientation lectures made no mention of the dioxin, lead, and arsenic contamination that could put their health at risk.

## RADIOACTIVE CONTAMINATION ON TORI SHIMA

When governments manufacture warheads for atomic weapons and fuel rods for nuclear power stations, they create as a byproduct large volumes of depleted uranium (DU). A heavy metal with toxic features similar to mercury and lead, DU's dangers are further heightened by the fact that it is radioactive.

DU is favored by the military for two main reasons. First, it can be used to make armor-piercing rounds, called penetrators, which unlike other metals that flatten on impact, retain their sharp points when they strike reinforced concrete or the outer shell of a tank. Second, DU is economic. Because DU is a waste product from the nuclear industry, the US government provides it to the military at very low cost or for free. The Pentagon first made extensive use of DU rounds in Iraq and Kuwait in 1991, then in the Balkans throughout the 1990s and again in Iraq in 2003. Most recently, despite its own promises not to do so, it also fired DU rounds against Islamic State forces in Syria in 2015.

When DU ordnance impacts a target, it produces a cloud of dust that spreads over a wide area and remains radioactive for a long time. Medical professionals in the places where DU was fired have cataloged the health toll of these munitions. In Basra, Iraq, for example, childhood leukemia rates are reported to have doubled between 1993 and 2007. Meanwhile, in Fallujah, there has been a spike in congenital malformations, which, in 2013, was estimated at about one in seven babies. Many environmental groups have called for DU to be classified as a weapon of mass destruction due to the long-term contamination it causes. Since 2007, the United Nations General Assembly has urged for investigations into its usage and, more recently, cleanup assistance for places where it has been fired. In much the same way as it handled Agent Orange, the US government has hindered DU investigations by refusing to release maps of where it has been used and downplaying the risks it poses to human health. Some scientists dispute claims that DU is the cause of health issues in the Middle East.[17]

In 1995 and 1996, the USMC contaminated Tori Shima, a small Okinawan island used as a target range, with DU. Previously unreleased documents obtained via FOIA revealed how USMC fighter jets had fired 1,520 "25mm DU armor piercing/incendiary penetrators" at the island in December 1995 and January 1996. Soon thereafter, the military took the unusual step of trying to clear them up. Although the FOIA documents do not explain why the USMC made the decision, it appears it may have fired the DU ordnance in error. Given the radioactivity of the rounds, the military on Okinawa needed permission from the US Nuclear Regulatory Commission to clear them away, and, in March 1996, Kadena Air Base received the requisite radioactive material license. Between March and April 1996, the Department of Defense surveyed the island, but they were only able to collect 192 penetrators—approximately 13 percent of the total fired. It conducted several follow-up trips to Tori Shima but ceased recovery efforts in August.[18]

At the time, none of these operations was made public. The documents suggest one reason for the delay was that DU was not listed in the Japan Environmental Governing Standards (JEGS) as a substance that requires reporting. But the real reason for the military's reluctance to publicize the incident was more likely due to its timing. In September 1995, three service members had raped a twelve-year-old Okinawan girl, sparking massive anger on the island. The perpetrators were convicted and imprisoned in March 1996, and the crime had forced Tokyo and Washington into the creation of SACO. If the revelations about Tori Shima had been made public during this turbulent time, the two governments might have been forced to offer more concessions and land returns on Okinawa. As it was, the US military postponed admitting what had happened until after the December 1996 publication of the final SACO report. In January 1997, it informed the Japanese government of the usage of DU at Tori Shima, which it described as a "mistake," and the news became public.

Following the announcement, Japanese government teams, accompanied by US military personnel, made several trips to the island, but they could only recover another fifty-five penetrators. The US military decided to categorize the remaining 1,273 rounds as "lost"; each of these rounds contained 148 grams of DU. Thus, in total, 188 kilograms of DU remained on or around the island.

The report stated these shells are comprised of three types of uranium—uranium-234 (which has a half-life of 247,000 years), uranium-235 (half-life of 710 million years), and uranium-238 (which comprises 99.75 percent of the DU shells and has a half-life of 4.5 billion years). According to the US military report dated 1999, the DU remaining at Tori Shima posed no health risk and so no further cleanup activities were necessary; however, the assurance is contradicted by more recent USAF records that, as of September 2010, revealed fears of potential exposure to DU were preventing the military from conducting ecological surveys of the island.[19]

Following the Tori Shima incident, the US Embassy promised the Japanese Ministry of Foreign Affairs that all DU ordnance had been removed from Japan; however, the media discovered approximately 400,000 DU rounds had been stored at Kadena Air Base in 2001. Today, the range at Tori Shima is still active, and every bomb dropped runs the risk of kicking up radioactive particles. Okinawan TV reports showed one exercise in May 2014, involving an undisclosed weapon that created a massive, dusty cloud visible from thirty kilometers away.[20]

## CONTAMINATION AT KADENA AIR BASE

With two 3.7-kilometer runways, 1,000 industrial buildings, and dozens of tanks storing a combined 215 million liters of fuel, Kadena Air Base is the largest USAF installation in Asia. It hosts the 18th Wing, the biggest combat wing in the USAF, and has launched wars in Korea, Vietnam, and the Middle East. More than 20,000 American service members, contractors, and their families live or work on the base, and it employs about 2,750 Japanese. Attached to the air base is the vast Kadena Ammunition Depot, once known as Chibana Army Ammunition Depot, where nuclear and chemical weapons had been stored prior to 1972.

The two bases play prime roles in the island's ecosystem. Kadena Air Base includes more than twenty groundwater wells, and its western edge abuts the sea; the Zukeyama Dam (now named Kurashiki Dam) lies next to the ammunition depot. Several rivers flow near the bases, feeding a purification plant at Chatan, the provider of drinking water for hundreds of thousands of people in seven municipalities, one of them being Naha City, the prefecture's capital. Despite this environmental importance, the USAF has never voluntarily made public these bases' impact on nearby communities. Only via FOIA and whistleblowers willing to risk potential dismissal—and even imprisonment—are we able to grasp the damage caused.

As of 2018, the USAF had released to me, via FOIA, approximately eleven thousand pages of accident reports, environmental investigations, and e-mails related to contamination at Kadena Air Base. Startling in these documents is the USAF's ignorance about how hazardous substances used to be handled there. Recorded are numerous incidents where service members stumbled upon pollution caused by their predecessors. Underground discoveries include contamination from petroleum, oil, and lubricants, and forgotten fuel storage tanks. In July 2014, the discovery of a buried barrel of unknown chemicals triggered e-mails urging responders to keep a "low profile, please. Don't want this release [sic] to press." Another report described conditions at a former ammunition storage area: "Military debris is scattered in almost every ravine . . . some ravines contain rusting drums, practice bombs, and occasional unexploded ordnance."[21]

The FOIA-released reports reveal the exposure of US and Japanese nationals to dangerous levels of lead and other heavy metals. For decades, an industrial-scale furnace at the depot had been burning ammunition and "other exotic pyrotechnics" without any emission controls. In 1993, investigators discovered this incineration had contaminated nearby land with extremely high levels of lead; the report stated there were small farms and vegetable plots in the area, and the site was near a waterway. Another burn pit, cited in an April 1994 report, was blamed for further lead concentrations in soil; again, farmers' fields were in the vicinity. When I asked the military whether it had warned local farmers about this contamination, a spokesman said it did not keep records of such notifications.[22]

Other heavy metal contamination was revealed in a 1997 internal report about the disposal of waste from sandblasting, which contained high levels of cadmium and chromium. Cadmium softens bones and causes kidney failure; chromium is highly carcinogenic. The report cited the risk that USAF members may have been exposed to these toxins and that this waste had formerly been buried on the base.[23]

In addition to heavy metals, Kadena Air Base surveys revealed serious contamination from asbestos in buildings including dormitories, dining halls, and boiler rooms. One of the locations was an abandoned hospital that had been used by service members for combat training. Their exercises involved using axes and chainsaws to breach asbestos-packed doors, resulting in the spread of powdery asbestos across a wide indoor area.[24]

The World Health Organization (WHO) estimates that asbestos is responsible for half of occupational cancer fatalities worldwide. In recent years, Japanese base employees have struggled to win compensation from Tokyo for illnesses attributed to their work in asbestos-contaminated environments. In 2014, Tokyo finally recognized that twenty-eight victims had been exposed, but survivor support groups and base worker unions estimate the number of sick is likely much higher.[25]

In April 2016, I sent the Kadena asbestos report to USFJ so it could alert its own service members who might have been exposed during training. But according to my inside sources, the military ignored my efforts and took no action to contact those potentially poisoned, a further sign that the military on Okinawa would rather sacrifice the health of its own service members than admit its bases are contaminated.

Such criminal negligence extends to on-base housing and schools. In September 2014, the inspector general of the Department of Defense published a report slamming the US military on Okinawa, including Kadena Air Base, for conditions in on-base housing. The report cataloged serious problems, particularly elevated radon readings and mold, "which could re-

sult in chronic health conditions." Similar conditions were also identified at
USMC housing on the island.[26]

In August 2015, the USAF also released a report on the safety of the
water in its eight educational institutions on Kadena Air Base. The investigation discovered that 165 taps were supplying water contaminated with
lead above the Environmental Protection Agency's (EPA) safe level of 20
parts per billion (ppb). Some of these dangerous taps were in school kitch-
ens and drinking fountains; for example, one drinking fountain at Kadena
High School was contaminated with lead at 190 ppb. The report stated,
"Potential health effects include damage to the brain, red blood cells, and
kidneys. Other effects include low IQ, hearing impairment, and reduced
attention span." Among the schools with lead-contaminated water were
the two located next to the 108-barrel dioxin dumpsite: Bob Hope Primary
School and Amelia Earhart Intermediate School.[27]

## PCBS: A CAN OF WORMS

In the twentieth century, PCBs were commonly used as coolants in electrical transformers, but their manufacture was banned in 1979, due to
increasing evidence of their health risks. Today they are categorized as
persistent organic pollutants, which damage the nervous, immune, and
reproductive systems, and are linked to cancers. PCBs do not deteriorate
in the environment and can remain in the soil for many decades, and they
can bioaccumulate in the bodies of people who consume contaminated
food like shellfish and fish.

During the 1970s, service members stored large volumes of PCB-
containing oil in an outdoor pool measuring twenty-one meters wide at
Kadena Air Base, from where it was either sold to Okinawans or mixed
with fuel and burned. The existence of the pool only became known in
1998, when a former base worker told the Okinawa media about it and the
subsequent news reports prompted an official investigation. The storage
pool was located on a hilltop above Kadena Marina, a popular recreation
spot for swimming and fishing. In the past, tests had revealed PCBs in the
sea, suggesting contamination had spread from the pond via the groundwater or storm drains.

In February 2014, more information about PCB contamination at
Kadena Air Base came to light when I was contacted by Bob McCarty, who
had worked as deputy chief of public affairs at Kadena Air Base during the
1980s, and retained some internal reports that had raised his suspicions.

According to the documents, in November 1986, base officials discovered exceedingly high PCB pollution following a leak of transformer oil. Environmental tests revealed the spilled oil contained PCBs at a concentration of 214 parts per million (ppm), but the ground beneath it was contaminated at levels as high as 5,535 ppm. Both readings far exceeded international safety levels. "It appeared that the incident (the November 1986 spill) merely 'opened a can of worms.' Soil sampling in the open storage yard would have yielded high PCB levels whether the spill had occurred or not," stated the report.[28]

The whistleblower's reports once again show the US military was more concerned with public image than safety. The USAF worried that news of the contamination might damage the standing of conservative governor Nishime Junji, a supporter of the US military presence on Okinawa, in the run-up to prefectural assembly elections. The report also cited unease that if Kadena's PCB contamination was made public, it would lead to demands for tests on other US bases—something the military was desperate to avoid.

In 2000, the issue of how to dispose of PCBs from Kadena and other military bases in Japan became an international problem. During this period, the amount of PCB-contaminated equipment owned by the US military in Japan was approximately 3,220 tons. With much of this waste stockpiled at the US Army's Sagami General Depot, Kanagawa Prefecture, there were plans to transport it overseas for disposal. In April, the private container vessel *Wan He*, carrying approximately one hundred tons of the waste, attempted to dock in Canada, but it was turned away due to regulations prohibiting the import of hazardous waste. For the same reason, it was rejected from docking in the United States, so it temporarily returned to Japan.

The ship eventually offloaded its toxic cargo at Wake Island, a US territory located approximately thirty-two hundred kilometers southeast of Tokyo. The island was chosen because of its grey zone status, where US customs laws did not apply, a technique frequently used by the military to do things it can't do elsewhere on remote islands far enough from the US mainland to avoid civilian oversight. Indigenous rights groups slammed the decision, stating, "The use of the 'vast empty spaces' of the Pacific ignores the importance of land and waters throughout the region as a source both of sustenance and identity for indigenous peoples." An editorial in the *Honolulu Advertiser* also questioned the wisdom of storing toxic waste on a low-lying island susceptible to inundation by storms.[29]

On Okinawa, Kadena Air Base announced its PCB cleanup was completed in June 1992, but problems still plague the base. According to FOIA-obtained reports, in 2011, internal investigators categorized the base's policies regarding PCBs as a "major deficiency." They underlined the absence of a safe storage area for contaminated transformers and failures to label equipment deemed hazardous. According to the same report, the installation had approximately five hundred transformers, but checks for PCBs had been conducted on less than half of them; in recent years, these transformers have leaked and, in some cases, even exploded.[30]

## DANGERS FROM THE AIR

FOIA-released reports reveal numerous instances where military aircraft dumped fuel over civilian communities, exposing people on the ground to benzene, naphthalene, and toxic additives linked to serious illnesses. Because JEGS does not apply to US military aircraft, USFJ does not need to curtail or report such contamination.[31]

People living close to Kadena Air Base—and other installations, including MCAS Futenma and Atsugi and Yokota air bases on mainland Japan—have long maintained that aircraft noise was damaging their health. The connection between such long-term exposure to noise pollution and serious illnesses has been well-documented. For example, a 2011 WHO report showed it can cause heart problems and tinnitus, and damage sleep patterns, resulting in further troubles, like poor memory consolidation. Such exposure is especially harmful to children, whose cognition skills can be damaged.[32] Research by one Japanese university estimated that noise from Kadena Air Base is responsible for an average of ten deaths a year, and approximately seventeen thousand people in nearby communities suffer from disturbed sleep.[33]

The Japanese government makes token efforts to reduce the problem by providing some homes near Kadena Air Base with soundproofing, but the problem persists. Japanese courts have repeatedly recognized the harm noise pollution causes to those living nearby. In 2017, about twenty-two thousand residents were awarded a total of about 30.2 billion yen ($274.5 million)—the largest-ever judgment against the government of Japan. In his decision, the judge ruled that Kadena aircraft noise damaged sleep and caused an increased risk of high blood pressure; he also noted that the impact on children was particularly severe, and the noise might trigger traumatic memories for Battle of Okinawa survivors.[34]

Just as the US military is allowed to wash its hands of the responsibility of dealing with other types of contamination, the same has occurred with its noise pollution, whereby court-ordered damages are footed by Japanese taxpayers on the grounds that the US military is acting on facilities provided by the Japanese government in the interest of the Treaty of Mutual Cooperation and Security between the United States and Japan.

## LEAKS, SPILLS, AND THE IMPACT ON LOCAL WATER

On Kadena Air Base, there are twenty-three wells, some of which contribute to both on- and off-base drinking water, and more than three hundred thousand meters of drains carry storm water from the installation into local rivers. Ever since water from fuel-soaked wells ignited in the late 1960s, people have worried about the safety of the local supply, but it was not until the release of FOIA documents that it became evident just how badly the base had been polluting this system.

One of the reports, dated August 1992, focused on contamination from the base's firefighting practice area, a site situated on a hilltop above Kadena Marina, near where the PCB storage pool had been located. Investigators discovered that fuel and firefighting foam, both hazardous substances, were flowing from this training area into the sea. The military, according to the report, did not know if this was in violation of Japanese environmental laws because the only available translations were not clear; however, the discharge was against the military's own rules, which banned the discharge of foam into storm drains.[35]

The second area of concern was a corrosion control area where aircraft were washed, their paint was stripped with chemicals, and they were repainted. Some of the effluent from this area soaked into the ground and flowed into the sea. Contaminants, including boron, phenol, heavy metals, and cyanide, were discovered in the runoff.[36]

To this day, environmental safety at Kadena Air Base remains poor. FOIA-released documents dated from 1998 to 2016, catalog approximately 650 environmental incidents at the installation, with 253 occurring since 2010. These incidents have ranged from small fuel leaks of only several liters that stayed on the base to large spills discharging tens of thousands of liters of fuel and raw sewage into local rivers. As with the USMC accidents, the Kadena reports show military incompetence has repeatedly endangered local water.[37]

In June 2013, an overflowing manhole leaked 208,000 liters of sewage into the Hija River, but the base took twenty-seven hours to notify local authorities. In-house e-mails included the comment, "We received little media coverage. So that's good news." Likewise, in July 2014, service members dumped hundreds of liters of medical waste—described as "expired shelf-life injectable fluids"—into on-base drains. "It's very unlikely that anything will be seen or reported, but if the milky solution reached the Hija River we will have a very upset public," stated the report. The accident was not reported to the Japanese government.[38]

## PFAS: THE TWENTY-FIRST CENTURY AGENT ORANGE

Fires are a constant fear at all airports but even more so at military ones, where aircraft, in addition to fuel, also carry high explosives. To extinguish fuel fires, water is useless—at best it merely evaporates, but in a worst-case scenario it can cause a flash explosion similar to pouring water on a pan of burning cooking oil. In the 1960s, the US Navy helped develop a special foam that could smother fuel fires and quickly extinguish them; they named it aqueous film-forming foam (AFFF). Thousands of military and commercial airfields installed the substance in fire trucks and hangar sprinkler systems. Large volumes of the foam were also sprayed during practice drills at fire training pits, including the one at Kadena Air Base, which emptied into the sea.

In a history frighteningly reminiscent to that of its rainbow herbicides, for decades the US military covered up evidence of the toxicity of AFFF ingredients, per- and polyfluoroalkyl substances (PFAS). USAF research from 1979, revealed these chemicals damaged the cells and liver of laboratory animals and caused low birth weight. Further reports in the 1980s identified the poisonous effects of AFFF but the military never made public the dangers, and on its bases throughout the world, it continued to use the foam, spraying millions of liters in training exercises. Meanwhile, the manufacturers 3M and DuPont, who used PFAS for nonstick frying pans and water-repellent clothing, had also been aware of the substances' risks but kept them quiet.[39] Due to these cover-ups, it took decades for US authorities to become aware of the dangers. Only in 2006, for example, did the EPA issue its first warning that the foam might be carcinogenic—but still the military kept spraying it.

Today, the EPA recognizes the myriad of serious health problems caused by PFAS: "developmental effects to fetuses during pregnancy or to breast-fed infants (e.g., low birth weight, accelerated puberty, skeletal variations), cancer (e.g., testicular, kidney), liver effects (e.g., tissue damage), immune effects (e.g., antibody production and immunity), thyroid effects, and other effects (e.g., cholesterol changes)." What makes PFAS so dangerous is their persistence. In the human body, they have a half-life of five and a half years, and in the environment, they remain toxic for long periods, too, earning them the nickname "forever chemicals." Because the US military has been spraying AFFF for decades, it has entered the environment near its bases; in the United States, AFFF use from approximately three hundred military sites has poisoned nearby groundwater.[40]

In 2016, the EPA set a drinking water advisory level for the PFAS compounds perfluorooctane sulfonate (PFOS) and perfluorooctanoic acid (PFOA) at 70 parts per trillion (ppt). Although the number might sound infinitesimal, the foam is so toxic that only a few drops in an Olympic-sized swimming pool would pollute the entire volume, exceeding the EPA level. Many experts believe even 70 ppt is too high, and some states have introduced lower advisories, for instance, Vermont, where the level is 20 ppt. In 2018, the White House halted the release of a report from the Agency for Toxic Substances and Disease Registry to avoid a "public relations nightmare"—the report showed PFAS were more toxic than previously revealed.[41]

Documents obtained via FOIA reveal that the US military's cover-up of the dangers of its firefighting foams extended to Japan, and frequent accidents have contaminated water sources on Okinawa. The reports show that USFJ was aware of their potential dangers from at least as early as 1992, when investigators criticized Kadena Air Base for allowing their release from the training pit into the sea. Between 2001 and 2015, the base mistakenly released at least twenty-three thousand liters of various firefighting foams, including those containing PFAS. Photographs of one incident in December 2013, show the aftermath of a spill—blamed on a technical malfunction—that poured out of an open hangar and into storm drains. The images of cars buried to their windows and billowing bubbles are eerily beautiful—as though snow has fallen on subtropical Okinawa. But the illusion is quickly shattered by the realization these chemicals are hazardous to human health.

In May 2015, a drunk US marine trespassed into a hangar on Kadena Air Base and triggered the sprinklers, releasing 1,510 liters. The US military described the incident as an act of vandalism. Even though the spill affected local water, the SOFA ensured he was not punished under Japanese law—and the incident apparently was not reported to the Japanese government.[42]

As well as these accidental releases, huge volumes of AFFF have also been sprayed during routine fire-training exercises. Kadena is not the only US military base on Okinawa where PFAS spills have occurred. At MCAS

*Since 2001, accidents on Kadena Air Base have released tens of thousands of liters of firefighting foam, a substance linked to water contamination. These photographs show an incident blamed on a malfunction that discharged twenty-three hundred liters of foam in December 2013.* Obtained via the Freedom of Information Act

Futenma, between 2005 and 2009, there were at least three spills involving a total of 2,669 liters of AFFF. One incident, in August 2007, leaked off the base "into a short creek, then immediately into a cave." FOIA-released reports from MCAS Iwakuni, Yamaguchi Prefecture, also reveal that the USMC in Japan has long been aware of the dangers of its firefighting foams—but it continued to use them.[43]

Apparently, the US military concealed from the Japanese government information about the toxicity of AFFF for almost four decades. The US military first became aware of the dangers in 1979, but it was not until April 2016, that PFOS was added to the list of reportable hazardous substances under JEGS, and as of January 2020, PFOA had not yet been added. This prolonged cover-up has hurt Okinawans; the military's use of PFAS has contaminated Okinawa's drinking water at concentrations exceeding the EPA's advisory limit of 70 ppt. In 2008, levels in a well on Kadena Air Base measured as high as 1,870 ppt; between February 2014 and November 2015, checks on the Dakujaku River near Kadena Air Base peaked at 1,320 ppt.

Most worrying was the discovery of contamination at Chatan Water Purification Plant, which supplies drinking water to seven municipalities in Central and South Okinawa, including Naha. In 2015, combined PFOS and PFOA levels recorded there reached 120 ppt, exceeding the EPA guidelines. To lower the concentration, the Japanese government installed activated carbon filters at an annual cost of 170 million yen ($1.5 million); however, they do not remove all PFAS from the treated water.

Tests near MCAS Futenma have also revealed numerous water sources, including wells and springs, contaminated at levels above the EPA limit. In 2019, blood tests on residents living near the base revealed their PFOS levels were four times the national average, and levels of the related compound perfluorohexane sulfonate (PFHxS) were fifty-three times the national average. Farmers' fields that are fed by water running beneath MCAS Futenma were contaminated with PFOS at 11,436 ng/kg, higher than any level ever detected by the Japanese government.[44]

The US military's reaction to the discovery of this contamination on Okinawa has been characteristically hypocritical; in the United States and overseas, it has cooperated with polluted communities, but on Okinawa, it has done nothing. For example, in March 2016, the USAF promised to conduct tests for PFAS contamination on 664 bases in the United States—and also US Army Garrison Ansbach, Germany—and they provided contaminated communities with alternative water supplies like bottled water. In October 2017, the Department of Defense revealed it had also discovered PFAS contamination at four US Army bases in South Korea. As a result, it isolated and shut down several wells. On Okinawa, the US military has offered no such support.

Despite the 2015 Cooperation Concerning Environmental Matters, the military has not granted permission for on-base checks, and in response to Okinawa Defense Bureau requests to the USMC, a spokesperson wrote, "There is no point in responding to additional questions or holding a meeting for which there are no established standards nor regulations." In June 2019, during questioning in the parliament, the Japanese government let slip that it actually possessed the US military's PFAS contamination data—but it couldn't share the information without the military's consent. The admission angered many Okinawans.[45]

Once again, FOIA requests were able to penetrate some of this joint US–Japan opacity. Released documents revealed Kadena Air Base was using extremely contaminated AFFF in at least eleven sprinkler units and the fire training pit near Kadena Marina was polluted. Likewise, the pit on MCAS Futenma was contaminated with PFOS at 27,000 ppt. The site is located uphill from the badly contaminated springs and farmers' fields.[46]

FOIA requests also uncovered how the USMC had disposed of waste AFFF on Okinawa in recent years. Between 2014 and 2015, it had contracted a local agency to incinerate and then bury at least 142 tons of firefighting chemicals, which the military had told the company were "waste alkali." Subsequently, high levels of PFAS contamination were detected in groundwater near the landfill site where the chemicals were dumped.[47]

The contamination of the island's water system with PFAS is a public health crisis unprecedented even in the US military's appalling environmental record on Okinawa. Its actions have poisoned the drinking water supply of not only the communities around the bases, but also those many kilometers away. This pollution has likely been going on for decades, and people have been consuming contaminated water without their knowledge. Today, this pollution jeopardizes the health of hundreds of thousands of Okinawan residents and the health of its own service members—not to mention the health of the millions of tourists who visit the island each year.

Despite Okinawans' demands for a comprehensive investigation, the Japanese government has not conducted its own checks near bases, nor has it introduced a national safety level for PFAS in drinking water. This neglect leaves residents uninformed, unprotected, and with no recourse for justice. Its refusal to share the military's on-base test data follows a long pattern of complicity, prioritizing the US military over its duty to protect its own citizens. Such complacency is as criminal as the actions of the US military itself.

# Timeline

# ENVIRONMENTAL CONTAMINATION AND ACCIDENTS ON OKINAWA (1947–2019)

| | |
|---|---|
| 1947 | On Iheya Island, eight villagers die from arsenic poisoning linked to cannisters dumped by the US military near a water well. |
| 1953 | The US Army starts to bring chemical weapons to Okinawa; by 1965, thirteen thousand tons of mustard, sarin, and VX agents are being stored at Chibana Army Ammunition Depot. |
| October 1955 | Children are cut by flying glass following first test of a nuclear cannon near Matsuda Elementary School in Ginoza Village. |
| May/June 1957 | A large leak from an army fuel tank damages farmers' fields in Gushikawa (current-day Uruma City). |
| June 1959 | At Naha Port, a misfired Nike nuclear missile kills two technicians; the missile lands in the sea and is later recovered in secret. |
| May 1961 and September 1962 | Military scientists conduct eleven tests of biological weapons (rice blast) at Shuri (Naha City), Ishikawa, and Nago City. |
| October 1964 | A US Army photograph captures gas-masked soldiers dumping unidentified chemicals into Okinawa's sea. |
| March 1965 | At Ginoza Middle School, students are sickened by CS gas from a nearby training area. |
| December 1965 | The US Navy loses a one-megaton hydrogen bomb while transporting it from Southeast Asia to Yokosuka Naval Base aboard the aircraft carrier USS *Ticonderoga*; it still lies northeast from Okinawa's coast. |
| 1967–1968 | Civilian wells near Kadena Air Base are so polluted with fuel that water drawn from them can catch fire. |
| January 1968 | A fuel pipeline breaks on Marine Corps Air Station (MCAS) Futenma, swamping 165,000 square meters of fields and rice paddies, and contaminating the water source for 280 households. |

| | |
|---|---|
| July 1968 | Two hundred thirty children are burned while swimming near Gushikawa; US forces had been spraying defoliants nearby at the time. |
| August 1968 | At Naha military port, there is a suspected leak of radioactive cobalt-60. In May 1972, cobalt-60 is again detected there and in shellfish at White Beach. |
| July 1969 | A leak of sarin agent at Chibana Army Ammunition Depot injures twenty-three US service members and one civilian. According to veterans, another leak sickens three soldiers in 1971. Operation Red Hat removes chemical weapons from Okinawa in 1971. |
| 1971 | Surplus military herbicides contaminate civilian water supplies in Haebaru and Gushikami districts. |
| 1972 | Two leaks of CS gas at Chibana Army Ammunition Depot injure several American service members and Okinawan base workers; in January 1973, more CS gas seeps into Yomitan High School, injuring students. |
| February 1973 | At Makiminato Service Area, a spill of anticorrosion compound forces the evacuation of 180 workers; symptoms include irritated eyes and throats. |
| September 1973 | On Iejima, US forces spray defoliants on antibase farmers' fields, affecting two thousand square meters. |
| December 1974 | At Makiminato Service Area, leaks of Vietnam War surplus chemicals kill large numbers of fish; similar deaths occur in January 1975 and October 1986. An army investigation blames the contamination on pesticides, dioxin, and polychlorinated biphenyls (PCBs). |
| June 1975 | Camp Schwab spills sewage into sea, damaging fishermen's nets. |
| August 1975 | A large leak of industrial detergent exposes workers to hexavalent chromium and other toxins at Makiminato Service Area. |
| January–February 1976 | The US Army buries hazardous waste on Makiminato Service Area and pesticides at Camp Hansen. |
| 1976 | Large spills occur throughout the island. In January, a sixteen-thousand-liter diesel leak contaminates the Kokuba River; two spills of diesel in January and June damage the shoreline at Ginowan; in September, service members spill oil and detergents into Tengan River, damaging farmers' fields. |
| March 1980 | Patients and staff from Ryukyu Psychiatric Hospital, as well as local residents, are sickened by a tear gas leak at Camp Hansen. |
| Summer 1981 | US marines unearth more than one hundred barrels suspected of containing defoliants on MCAS Futenma; top officers cover up the incident. |
| 1987 | The United States Air Force (USAF) discovers severe PCB contamination at Kadena Air Base. |
| June 1988 | Two civilians driving near the Northern Training Area (NTA) fall ill after exposure to CS gas. |
| August 1992 | At Kadena Air Base, an internal report reveals that firefighting foam is leaking off the base. |

| 1993 | A USAF report reveals that a munitions incinerator on Kadena Air Base and another burn pit have been contaminating fields with lead. |
| December 1995–January 1996 | United States Marine Corps (USMC) jets fire 1,520 depleted uranium shells at Tori Shima range; subsequent cleanup operations only retrieve 16 percent of the ordnance. |
| 1996 | Land returned to civilian use on the former Onna Communication Site is found to contain dangerous levels of mercury, cadmium, and PCBs. |
| 1999 | On Camp Courtney, lead contamination at a skeet-shooting range raises fears of damage to nearby areas growing hijiki. |
| 2002 | One hundred eighty-seven barrels containing an unidentified, tar-like substance are unearthed on former military land in Chatan. |
| 2003 | High levels of lead and hexavalent chromium are discovered on land in Chatan that was previously part of Camp Kuwae. |
| September 2005 | At Camp Schwab, contractors accidentally cut a fuel line, contaminating 120 meters of river. |
| May 2007 | Fifteen thousand liters of jet fuel leak on Kadena Air Base. |
| August 2007 | Seven hundred fifty-seven liters of firefighting foam spills on MCAS Futenma, some of which leaks off base "into a short creek, then immediately into a cave." |
| November 2008 | At Camp Hansen, a marine washes unknown chemicals into drains which then flow off the base near an elementary school; in May 2010, antifreeze leaks from the base into the sea. |
| November 2010 | From Kadena Air Base, fifty-seven thousand liters of raw sewage contaminate the Shirahi River and the sea. |
| March 2012 | An abandoned underground tank leaks hundreds of liters of fuel at Kadena Ammunition Depot. |
| February 2013 | At Camp Foster, a US marine causes a nineteen-thousand-liter sewage leak by pouring cooking oil down a sink; a USAF civil engineering officer is too afraid to reprimand him for fear of a violent reaction. |
| June 2013 | From Kadena Air Base, 208,000 liters of sewage leaks into Hija River, but the military takes twenty-seven hours to notify local authorities. One hundred eight barrels contaminated with dioxin, herbicides, PCBs, and other toxins are unearthed from a soccer pitch in Okinawa City. |
| August 2013 | At Camp Hansen, an HH-60 Pave Hawk helicopter crashes near Okawa Dam. Subsequent tests of the soil at the crash site by the US military reveal arsenic, cadmium, and lead. |
| December 2013 | At Kadena Air Base, 2,270 liters of firefighting foam spills from a sprinkler system and enters storm drains. |
| 2014–2016 | At Kadena Air Base, USAF checks on a pond reveal perfluorooctane sulfonate (PFOS) contamination of 90,000 parts per trillion (ppt), and checks on a sprinkler system reveal PFOS contamination of 9,500,000,000 ppt. |

| | |
|---|---|
| July 2014 | At Kadena Air Base, service members dump hundreds of liters of medical waste into on-base drains flowing to Hija River. |
| January 2015 | Near White Beach, a US Navy vessel dumps 150,000 liters of wastewater, but local authorities are not informed for three days. |
| May 2015 | At Kadena Air Base, a drunk US marine trespasses into a hangar and triggers the sprinkler system, releasing 1,510 liters of firefighting foam. |
| January 2016 | The discovery of PFOS contamination at Chatan Water Treatment Plant is announced; the plant supplies water to approximately 450,000 residents and tens of thousands of service members and tourists. |
| February 2016 | At MCAS Futenma, USMC checks on a firefighting training pit reveal PFOS contamination of 27,000 ppt and (perfluorooctanoic acid) PFOA contamination of 1,800 ppt. |
| June 2016 | At MCAS Futenma, an accident spills 6,908 liters of aviation fuel. |
| August 2017 | A 375-liter leak of firefighting foam contaminated with PFOA occurs at Kadena Air Base, with some entering storm drains and blowing off base. |
| October 2017 | A USMC CH35 helicopter crash-lands in Higashi Village. The Okinawa Defense Bureau discovers contamination from benzene and strontium-90. |
| May 2019 | Blood checks on Ginowan residents reveal high levels of PFOS, PFOA, and perfluorohexane sulfonate (PFHxS); the soil is also contaminated. |

# 7

# JAPAN

## Contamination, Nuclear Deals, and the Fukushima Meltdowns

Today, mainland Japan hosts forty-seven US bases. Yokota Air Base, Tokyo, is the headquarters of United States Forces in Japan (USFJ) and the gateway for visiting US presidents; its 3,300-meter runway serves the 374th Airlift Wing, and the base contains the National Security Agency (NSA) post where former CIA contractor Edward Snowden was assigned in 2009. Naval Air Facility Atsugi (NAFA), Kanagawa Prefecture, is the largest US Navy air base in the Pacific. The arrival point for General Douglas MacArthur in 1945, it went on to host CIA operations during the Cold War, and Lee Harvey Oswald once worked there as a radar controller.

Yokosuka Naval Base, Kanagawa Prefecture, is the Department of Defense's largest overseas port, home to the Seventh Fleet and the only US nuclear-powered aircraft carrier stationed outside the United States, the USS *Ronald Reagan*. Other large US military bases on mainland Japan include Camp Zama, Kanagawa Prefecture, the headquarters of the US Army in Japan, and Misawa Air Base, Aomori Prefecture, where the US Navy, Air Force, and Army share the installation, with a surveillance site used to spy on Asia and the rest of the world.

## CONTAMINATION IN THE COLD WAR ERA

Unlike on Okinawa, where bases were built on stolen land, many of the US facilities on mainland Japan used to belong to the Japanese Imperial military.

Yokosuka Naval Base, for example, was established in the 1860s as Japan's first modern arsenal, and the air base at Atsugi was established by the Imperial Japanese Navy in 1938. Early Japanese military operations contaminated these properties. Freedom of Information Act (FOIA)-released reports from the US Navy at Yokosuka, for instance, cite twenty-first-century discoveries of soil contamination blamed on fuel spills from pre-1945, and many of the bases had stockpiles of chemical weapons during World War II.[1]

Soon after Japan's surrender, residents living near US military bases became aware of the environmental problems caused by the new overseers. At Tachikawa Air Base, Tokyo, leaks from waste oil storage tanks and fuel pipelines contaminated the local water supplies; by 1952, almost ten thousand residents had been exposed, and just like Kadena Air Base, the contamination reached such high concentrations that water drawn from wells could be set alight. Among the substances detected in Tachikawa's water was tetraethyl lead, which can damage the central nervous system and cause reproductive disorders. Public anger at this contamination contributed to mass protests, known as the Sunagawa Struggle, against the installation and its proposed expansion in the latter half of the 1950s.[2]

In 1952, the Treaty of San Francisco ended the US occupation of mainland Japan, but ongoing accidents and crimes involving service members—particularly the United States Marine Corps (USMC)—infuriated local communities. In the following years, the US government relocated troops to Okinawa, where, with the entire island under its control, it was able to seize private land and trample human rights with impunity. Between 1955 and 1960, the number of military bases on the mainland dropped from 658 to 241; during the same period on Okinawa, the military presence almost doubled.[3]

By 1965, when the US military deployed its first combat troops to South Vietnam, there were about 150 installations on mainland Japan. In his authoritative book *Fire across the Sea* (1987), Thomas R. H. Havens details how the US military used these bases, purportedly in Japan to protect the nation, to wage aggression in Southeast Asia. The Vietnam War opened the eyes of many Japanese people to the realities of the Japan–US Treaty of Mutual Cooperation and Security, according to which fellow Asians were slaughtered by missions launched from Japan with the tacit support of Tokyo. Most Japanese people opposed the war, many staged huge demonstrations against it, and some families helped shelter US deserters. But at the same time, the Vietnam War—like the Korean War—was a boon to the economy; between 1966 and 1971, Japanese companies reaped approximately $1 billion a year.[4]

Just as the Pentagon had declared the United States couldn't fight in Vietnam without Okinawa, US ambassador Ural Alexis Johnson was equally explicit about the role of mainland Japanese bases. "Japan was vital to our effort in Vietnam," he said. "It provided ports, repair and rebuild facilities, supply dumps, stopover points for aircraft, and hospitals for badly wounded soldiers."[5]

These military operations polluted almost everything they touched.

To Sagami General Depot, the army brought war-damaged vehicles for repair. Such work involved the disposal of large volumes of battery acids, oils, and solvents. In 1972, the carcinogenic heavy metal cadmium was dumped in a local river, and in the following years, there were severe problems related to polychlorinated biphenyls (PCBs).[6]

Flights from mainland airfields transported troops and supplies to Okinawa and then the war zone. Tachikawa Air Base, for example, logged more than five hundred flights a week during the Vietnam War. Spills of fuel and oil contaminated the surrounding environment. NAFA possessed the US military's only full repair center for aircraft engines—a process involving copious volumes of carcinogenic solvents and anticorrosion chemicals. Veterans from Marine Corps Air Station (MCAS) Iwakuni recall spraying defoliants on the installation to kill vegetation, and according to Alvin "Doctor Orange" Young's own admissions, the military approved sending surplus stocks of arsenic-containing Agent Blue to Misawa Air Base in the early 1970s.[7]

Whereas on Okinawa fuel was pumped across the island via the military's network of leaky pipes, on the mainland it was transported by trains through urban areas. At its peak, an estimated 5 million liters of fuel a day passed through central Tokyo on its way to US military airfields in the western suburbs. One morning in January 1964, a tanker carrying military fuel collided with a stationary passenger train at Tachikawa Station; the subsequent blaze destroyed eight homes; two passengers were injured fleeing the scene. Disaster struck again three years later in the heart of Tokyo when, in August 1967, a freight train crashed into fuel tank cars at Shinjuku Station. The fire took hours to extinguish and led to the cancellation of more than a thousand trains.[8]

Mainland military ports, already suffering contamination by fuel, solvents, and sewage, faced a new threat beginning in 1964, when US nuclear-powered submarines started visiting Sasebo Naval Base, Nagasaki Prefecture, and then, two years later, Yokosuka. In 1968, the first US nuclear-powered aircraft carrier, the USS *Enterprise*, came to Japan. Tens of thousands of protesters gathered outside the bases, angry about the dangers of leaks from on-board reactors and suspicions—later proved justified—that they were carrying nuclear weapons.

US military officials assured local communities that such vessels were safe, but some incidents debunked such promises. In January 1969, the USS *Enterprise* experienced a fire while off the coast of Hawaii. Sailors had positioned a starting unit too close to rockets, and its hot exhaust caused an explosion that set off fires and other ordnance detonations, destroying 15 on-board jets, killing 28 crew members, and injuring another 314. The fire blazed for four hours but was extinguished before it could reach the ship's eight nuclear reactors or its store of an estimated one hundred nuclear bombs. Following the fire, the captain admitted the USS *Enterprise* had almost been destroyed.[9]

One year previously, in May 1968, radioactive contamination had been detected at Sasebo harbor near the nuclear-powered submarine USS *Swordfish*. When asked how the Japanese government would respond to a nuclear accident at the base, Prime Minister Sato Eisaku replied, "Not only is the population there small, but we don't even think about such things."[10] Combining equal measures of disdain for the communities burdened with bases and ignorance of his government's own responsibility to protect citizens' lives, Sato's comments epitomized Japanese authorities' attitude toward nuclear matters during this period.

## US NUCLEAR WEAPONS IN COLD WAR JAPAN

As with other US weapons of mass destruction, major obstacles impede research on the storage of nuclear weapons on mainland Japan; the Department of Defense maintains a blanket policy of neither confirm nor deny, and on the Japanese side, Tokyo feigns ignorance behind this US wall of silence; however, FOIA requests and interviews with whistleblowers and veterans have been shedding some light on these matters.

The earliest in-depth research was gathered by Nautilus Institute's East Asia Nuclear Policy Project in July 1999, and published as "Japan under the US Nuclear Umbrella"; more recently, Daniel Ellsberg's *The Doomsday Machine* (2017) revealed further disturbing information. From this research, two significant points become clear: 1) the wide extent to which the United States brought nuclear weapons to mainland Japan throughout the Cold War and 2) how successive generations of Japanese leaders were aware of these breaches of their nation's nonnuclear principles.[11]

According to Ellsberg, in the late 1950s, the US Navy anchored a ship, the USS *San Joaquin County*, containing nuclear weapons, approximately two hundred meters offshore from MCAS Iwakuni. The stationing of the vessel, which was disguised as an electronic repair ship, was the result of ghoulish interservice rivalry. The Navy wanted its marines to beat the Air Force to be the first to drop The Bomb if the order to go nuclear was given. Whereas it would take several hours to fly such weapons from Kadena Air Base to mainland United States Air Force (USAF) airfields in Operation High Gear, the Iwakuni marines could receive theirs in a matter of minutes. Drills involving the weapons took place along the beach at the USMC base, and the USS *San Joaquin* remained there until 1967.[12]

By the late 1950s, stated Nautilus, "US nuclear weapons were stored at three bases and routinely shipped through nine others in Japan." Between 1960 and 1963, nuclear weapons were flown into Japan, US veteran Earl Hubbard told the *Mainichi Shimbun* in 1972. He explained how B-43 bombs—the same as the one lost from the USS *Ticonderoga* in 1965—were brought from the United States to air bases at Johnson (today, Iruma Japan Self-Defense Force [JSDF] Air Base, Saitama Prefecture), Misawa, and Yokota bases; he also cited Kadena Air Base as a destination.[13] In addition to these USAF flights, research points to the US Navy regularly bringing nuclear weapons to mainland Japanese ports. The Japanese government claimed the United States would need prior consultation to transport weapons in such a way; however, records from the US side dispute that such a condition existed.[14]

In September 1974, retired admiral Gene La Rocque, former commander of numerous warships, told the US Congress Joint Committee on Nuclear Energy, "My experience . . . has been that any ship that is capable of carrying weapons, carries nuclear weapons. They do not offload then when they go into foreign ports such as Japan or other countries."[15] In Cold War accounts of transporting nuclear weapons to mainland Japan, the name of one base repeatedly appears: Yokosuka. In the early 1960s, US submarines docked at Yokosuka with nuclear Regulus cruise missiles. During the same period, the USS *Midway* also brought weapons there. When the USS *Midway* was being prepared to be homeported at Yokosuka in 1972, the US State Department recommended that, for diplomatic reasons, its nuclear weapons be removed. But the chief of naval operations refused, describing such an action as "operationally unacceptable." It is believed the USS *Midway* maintained its onboard nuclear arsenal, likely hundreds of bombs, during its entire homeporting at Yokosuka Naval Base from 1973 to 1991.[16]

One of the most disturbing revelations in *The Doomsday Machine* concerns who was authorized to launch a nuclear attack. Throughout the Cold War, the US government allowed the world to believe that only the president himself could give such an order; however, Ellsberg reveals that navy commanders in the West Pacific—of which Yokosuka Naval Base was a part—had the authority to launch a nuclear strike without any specific order from the White House. Possible circumstances included if the fleet lost communication with the president in times of war—but this delegation of nuclear decision-making created the risk of a commander going rogue for his own personal reasons.[17]

At the same time, Japanese governments assured the public that there were no nuclear weapons at US military ports, reasoning that because the Pentagon had not notified them of such arsenals, no caches existed. Tellingly, in the 1980s, when news broke of the USS *Ticonderoga* losing a nuclear bomb on the way to Japan, Tokyo did not push the US military for an explanation. Both the 1965 USS *Ticonderoga* accident and the January 1969 near-sinking of the USS *Enterprise* highlighted the inherent dangers of storing nuclear weapons afloat—a practice much more dangerous than keeping them on land.

During the 1980s and 1990s, other accidents close to Japan involved military vessels almost certainly packed with nuclear weapons. In April 1981, the US nuclear submarine USS *George Washington* (SSBN-598)—not to be confused with the aircraft carrier of the same name—collided with the Japanese merchant vessel *Nissho Maru* south of Sasebo. Two Japanese crew members were killed, and the navy was criticized for not doing more to help search for survivors. In March 1984, the USS *Kitty Hawk* struck a Soviet nuclear-powered submarine in the Sea of Japan. The US ship was believed to have been carrying tens of nuclear weapons, and the Soviet submarine was equipped with nuclear torpedoes. Fortunately, none of these was damaged in the collision, which, had it been worse, could have contaminated large swathes of the Sea of Japan.

One of the most worrisome accidents occurred in June 1990, when the USS *Midway* suffered two explosions while sailing off the coast of Japan. Sparked by a leak of jet fuel, the blasts killed three sailors and injured fifteen more. It was suspected that the ship was carrying nuclear weapons at the time, but the navy stuck to its familiar tactic of neither confirming nor denying such an arsenal.

In 1991, the US government began the slow process of removing nuclear weapons from navy ships in response to international pressure from countries with nuclear prohibitions for visiting military vessels, for instance, New

Zealand and Sweden; Japan did not join in the call for the ban. In July 1992, President George Bush announced that nuclear bombs had been removed from the navy's above-sea ships, but its submarines continue to carry nuclear weapons, with the current total estimated at about one thousand devices.

## CONTAMINATED LAND AND WATER

In the 1990s, as the US government forced its military to become more transparent about environmental transgressions at home, there was a rare report about problems on bases in Japan. In 1991, hearings of the Committee on Armed Services House of Representatives into Department of Defense Environmental Programs revealed that PCBs had been discovered at army installations in Japan, for example, Camp Zama and Sagami General Depot, and the groundwater had been contaminated with the solvent trichloroethylene (TCE). The hearings also revealed PCB problems at USAF bases listed as Yokota, Misawa, and Kadena; the worst trouble existed at Yokosuka Naval Base, which was described as having at least three waste sites and caves in which hazardous substances were stored.[18] The information revealed at the hearings was scant, but it was a rare admission of the US military's environmental problems in Japan.

Other reports from the 1990s also highlighted severe trouble at Yokosuka. In April 1993, the installation announced the presence of heavy metal contamination and also that PCB-polluted soil had been scattered in a landfill project there. FOIA-obtained documents reveal that a landslide on the base in the 1990s caused a fuel spill of more than fifty thousand liters, which entered the sea and took two years to clean up. In 1998, the Japanese Ministry of Defense discovered high levels of toxic substances at berth 12, designed for nuclear aircraft carriers—the pollution included arsenic, lead, and mercury. Following a collapse at the same berth in May 2000, a survey found large numbers of deformed fish, some with spinal deformities and tumors.[19]

Other FOIA-released documents reveal Yokosuka Naval Base struggling with past contamination. A report from October 2006, detailed construction crews discovering polluted soil and being told to rebury it, along with contaminated water. Moreover, in 2007, the environmental office at the base revealed that, in the preceding few years, it had lost track of the location of an entire pipeline still filled with oil running beneath the installation. Elsewhere on the mainland, there were major fuel spills in the 1990s and early 2000s totaling tens of thousands of liters at Yokota, Atsugi, and the Tsurumi Fuel Depot.[20]

## FOIA DOCUMENTS LIFT LID ON ACCIDENTS THROUGHOUT JAPAN

Internal reports obtained via FOIA requests reveal a pattern of incompetence, causing mass leaks on two of the US military's key mainland air bases: MCAS Iwakuni and NAFA. At Iwakuni, there were at least 344 environmental incidents between 2007 and 2016, spilling approximately twenty-five thousand liters of jet fuel. As explained in the previous chapter, FOIA-released documents from Iwakuni confirm that the US military in Japan has been aware of the environmental dangers of its firefighting foams for twenty years. One accident report from August 1997, stated that the foams are "harmful to the water," and another spill report from February 5, 2013, described firefighting foam as a "hazardous substance." Moreover, in May 2015, a foam spill was described as "PFOS [perfluorooctane sulfonate] contaminated"—but despite it leaking off the base, according to the report, the Japanese government was not informed.[21]

Atsugi Air Base, too, has experienced serious accidents. Between October 2009 and November 2016, FOIA-released documents cataloged at least fifty-three environmental accidents that spilled a combined more than two thousand liters of fuel, diesel, and sewage.[22] Meanwhile, in Aomori Prefecture, Misawa Air Base has suffered from the dangers of contamination and the inadequacies of the Status of Forces Agreement (SOFA). According to

*Workers struggle to stop a leak of jet fuel from escaping MCAS Iwakuni in January 2015.* Obtained via the Freedom of Information Act

FOIA-obtained reports, between 2012 and 2017, there were at least four large releases of firefighting foams, notably one in July 2012, that leaked into the water system used to irrigate rice fields. Photographs from the scene show large volumes of bubbles, but it seems no tests for per- and polyfluoroalkyl substances (PFAS) were conducted, and the USAF stated the foam was "not harmful to humans and agriculture."[23]

Misawa Air Base has also caused numerous problems for local fishermen. In 1992, a military aircraft dumped fuel into Lake Ogawara, famous for such freshwater produce as clams and smelt, and a mainstay of the local fishing population; the Japanese government had to pay out 8 million yen ($73,000) and help build a loading facility for them. Another accident occurred in May 2012, when an F-16 from Misawa Air Base dropped its two fuel tanks off-base near a bombing range; the incident was not reported to local authorities. Then, in early 2018, another USAF F-16 dumped its two fuel tanks after an engine fire. The 4.5-meter-long tanks landed in Lake Ogawara, so fishermen, suspecting their catch had been contaminated with fuel, had to throw out 385 kilograms of clams and halt fishing in the lake for one month.[24]

At Yokota Air Base, Tokyo, headquarters of USFJ, between 2012 and 2018, spills of hazardous substances topped at least ten thousand liters; some of the substances leaked off the base, but Japanese authorities were not informed. PFAS firefighting foam has contaminated the installation. Checks on the base's water system in 2016 revealed PFOS/PFOA levels of 35 parts per trillion (ppt); drawn from eleven wells located on the installation, the water is supplied to approximately 11,500 people. A report published in 2005, by a Japanese university, linked wastewater from the installation to PFOS contamination of the Tama River of 440 ppt, the highest concentration detected in the river. As well as accidental spills of firefighting foam, Yokota Air Base also has a fire training area where PFOS-contaminated foams were sprayed for decades.[25]

As bad as the conditions on these bases have been, at least they bothered to keep records of their environmental accidents; the same cannot be said of Sasebo Naval Base. In response to a FOIA request for the facility's documents, it responded, "No detailed records exist prior to 2014." When it did begin to keep records, photographs underlined sloppiness, with sailors working without gloves, applying incorrect cleanup pads, and wrongly positioning an oil boom, allowing spilled substances to spread in the sea from the scene of the leak.

## FIRES AND NOISE POLLUTION

Since the 1970s, there have been large fires on US bases on the Japanese mainland, but due to restricted access, it has often been difficult for Japanese emergency crews to deal with them, and US follow-up investigations have been unsatisfactory. At Tsurumi Fuel Depot in July 1979, lightning struck a storage tank, setting it ablaze. The fire burned for four and a half hours, during which time local fire crews were denied access to the facility. Following the fire, the Japanese and US government's explanation of how it occurred left many questions unanswered. Two years later, in October 1981, there was a large explosion, followed by a four-hour blaze at Koshiba POL Depot, in Kanagawa Prefecture. Seven residents were injured, and there were 463 counts of property damage—including windows shattered by the blast. It took the US military almost two years to announce the results of their investigation, at which point they admitted they were unable to find the cause of the explosion.[26]

What happened in August 2015, at Sagami General Depot, reinforces how little base officials know about the installations under their charge. At 12:45 a.m., multiple explosions rocked one of the army warehouses, sending jets of flames high into the sky. But military and local firefighters were not allowed to extinguish the blaze because no one knew what kind of chemicals were in the warehouse and whether water might have caused a catastrophic reaction. Left untreated, the fire continued to burn for six hours.[27] As with the previous fires, the US military investigation took more than a year to complete, and even then, they were unable to pin down its cause with any certainty.[28]

Noise pollution from US military aircraft—as well as JSDF flights—has troubled residents living near military bases on the mainland for decades. Since the 1970s, numerous lawsuits have been filed by those in the vicinity of Atsugi, Yokota, and Iwakuni air bases. In recent years, the courts have ordered the Japanese government to pay compensation to residents[29]; however, these victories have been illusory, resulting in no real reforms, as the courts refused to ban early morning and nocturnal flights by either the JSDF or US military. The military—whether American or Japanese—takes precedence over the health of the people it claims to protect.

## THE CIA AND THE PROMETHEUS OF JAPANESE NUCLEAR POWER

In December 1953, President Dwight D. Eisenhower gave his "Atoms for Peace" talk at the United Nations (UN) to sell the world on the civilian use of US nuclear power and distract growing criticism from its ever-expanding program of weapons tests. Given its destruction of Hiroshima and Nagasaki less than a decade earlier, the US government understood that Japan would be the perfect showcase for nuclear power stations—and it found ready supporters among the Japanese conservatives who it had allowed back into powerful positions following its Reverse Course.

Pivotal to US machinations for nuclear power was Shoriki Matsutaro. Prior to World War II, Shoriki had been a police chief in Tokyo, cracking down on public demonstrations and launching raids on universities. Then, in 1924, he purchased the struggling *Yomiuri Shimbun* and soon transformed it into the most influential newspaper in Japan via an unlikely editorial line of supporting Japanese colonialism in Asia and promoting baseball by arranging visits by such US teams as Babe Ruth's New York Yankees. In 1936, he helped create a single government-run news agency, Domei Tsushinsha, which endorsed militarism and censored any news critical of the authorities, particularly, as time went by, Japan's losing war. Following surrender, Shoriki was arrested by the United States; his file noted, "He ought to be regarded as one of the most evil influences in poisoning the public mind." But during the Reverse Course, the United States realized it could make good use of his media and political clout. Thus, after dismissing accusations against him as leftist scuttlebutt, it released Shoriki from prison.[30]

Throughout its 1945–1952 Occupation, the United States waged a multipronged propaganda campaign in Japan; on one hand, its censorship removed anti-US sentiments, and on the other hand, the US Information Agency churned out thousands of hours of radio shows and cinema reels to promulgate pro-US feeling. Worrying the end of the occupation would limit such operations, the CIA reached out to Shoriki—who they dubbed PODAM—and others to wage psychological warfare on the newly restored sovereign nation. In 1952, backed by the CIA and the American Council, Shoriki set up Japan's first commercial TV station, Nippon Television (NTV); it quickly became a massive success.[31]

With Shoriki on side, the United States now could influence both the *Yomiuri Shimbun* and NTV. It consolidated its power by pumping millions of dollars into the newly created Liberal Democratic Party (LDP) to promote US interests and block the spread of progressivism. The funding continued for decades, and, since 1955, the LDP has remained in power almost continually, transforming Japan into a de facto one-party state, riddled with nepotism, corruption, and subservience to Washington.[32]

One of the prime goals for the United States, Shoriki, and the LDP was the promotion of nuclear power; stripped of the empire that had once provided Japan with coal and oil, Japanese conservatives realized their resource-strapped nation needed a way to produce electricity; nuclear power held the extra appeal of providing plutonium to make the nuclear weapons for which many rightists hankered. The United States realized the sale of nuclear reactors to Japan would be both a financial boon and a public relations (PR) victory that might go a long way in dispelling its image of using Asians as "nuclear cannon fodder." Starting on January 1, 1954, the *Yomiuri* began a series of articles lauding the wonders of nuclear power, the first of which was titled "Finally, the Sun Has Been Captured."[33]

Three months later, the Bravo disaster and contamination of the *Lucky Dragon* derailed their propaganda campaign, causing a public backlash against nuclear technology. In secret memos, the CIA decried those who "rang the gongs of pollution by atomic waste, the Bikini incident, and Hiroshima, attempting to exercise the by now well-known Japanese terror of atomic radiation."[34]

Redoubling its efforts, the United States paid $2 million "condolence money" to the Japanese fishermen. Soon thereafter, the Japan Atomic Energy Commission (JAEC) was created, and Japan and the United States signed the Atomic Energy Agreement to allow shipments of enriched uranium to Japan. Shoriki, now a member of the first LDP cabinet, was appointed the inaugural president of JAEC. His CIA file notes he touted himself as "Prometheus who was bringing this fire to Japan." Meanwhile, to win over skeptics in academia, the US government lavished Japanese university professors with trips to the United States to sell them on the benefits of nuclear power.[35]

This was accompanied by extensive propaganda. NTV broadcasted specials promoting nuclear power; science fiction featuring nuclear-propelled rockets; and Disney's *Our Friend the Atom*, which had been funded by General Electric and the US Navy. On the ground, the *Yomiuri* sponsored a traveling exhibition that drew huge crowds. As a result of these sustained PR campaigns, Japanese views on nuclear power began to shift; in 1956, 70 percent of the people had negative views of the atom, but two years later this had dropped to 30 percent. Nowhere was this transformation more apparent than in Hiroshima. In 1956, the nuclear exhibition was held there, attract-ing 110,000 visitors; then, in 1958, a similar show extolling the virtues of the atom was staged at the newly opened peace museum. The extent to which the campaign had successfully created a schism in public/political rhetoric is evident in the fact that no Hiroshima mayor has ever spoken out against nuclear power despite unflinching denouncements of nuclear weaponry.

In the coming decades, fifty-four nuclear reactors were built in Japan. This rapid spread was facilitated by Japan's "Nuclear Village," a loose network of politicians, academics, corporations, and the media, working together to convince the public, and one another, that nuclear power was cheap and safe. LDP politicians received campaign funds from the industry, and upon their retirement, they parachuted into high-paid corporate jobs in the companies they had previously overseen. Pronuclear academics re-ceived research grants, while antinuclear scholars were kept out of the loop. Power companies ensured media support and/or silence via extravagant advertising budgets. At a local level, the communities where the nuclear power plants were built—many of which were in rural areas already suffer-ing from declining populations and economies—were awarded subsidies to host the facilities, and the plants provided jobs for residents.[36]

The close relationship between government agencies, corporations that built nuclear reactors, and electric power companies ensured that the nuclear power industry was largely allowed to regulate itself, putting profits ahead of safety. Inevitably, such self-policing and lack of outside oversight led to results similar to the US military: The industry cut corners, downplayed risks, and ignored worst-case scenarios; for decades in Japan, for instance, the operators of nuclear power plants falsified safety records.

In 1999, an accident at the Tokaimura plant, Ibaraki Prefecture, spread radiation more than 1.5 kilometers, forced 300,000 residents to shelter in-doors, and exposed more than 600 people, two of whom died. In its wake, only minor reforms were introduced. More significantly, operators ignored the risk of earthquakes, siting plants near fault lines and building them to ride out only relatively small tremors.[37] Such incidents made many Japanese people skeptical of the benefits of nuclear power, and communities filed lawsuits against the construction or operation of such plants; however, the courts unfailingly ruled against citizens in favor of the industry. The main-stream media, instead of siding with the public, failed to place nuclear busi-nesses under scrutiny, instead basking in utilities' advertising revenues and personal favors. Top Japanese media executives were on a nuclear industry–sponsored junket to China when the folly of packing so many reactors into one of the most tectonically active places on the planet became clear.

For centuries, the coastline of the Tohoku region has been pummeled by tsunamis triggered by earthquakes in the Pacific Ocean; some of these waves have been tens of meters tall and deluged communities many kilometers inland, killing thousands. Despite such vulnerability, the Tokyo Electric Power Company (TEPCO) chose to locate six of its reactors at its Daiichi plant in Fukushima Prefecture on the coastline. TEPCO ignored the threat of a large tsunami; initially, it designed the plants to survive a three-meter wave, then revised this to six meters; finally, it settled on a ten-meter sea wall. In 2009, TEPCO was warned the plant could be struck by a far taller tsunami, similar in scale to one that had struck in the ninth century, but the company rejected the scenario and failed to act. TEPCO's accident plan for the plant summed up its complacency, assessing the risk of an accident at the Fukushima site as "practically unthinkable."[38]

On the afternoon of March 11, 2011, a magnitude 9.0 earthquake struck off the coast of Tohoku. One of the most powerful tremors ever recorded, it lasted for three minutes, shifted Honshu almost three meters, and caused a tsunami; almost twenty thousand people died. At the Fukushima Daiichi plant, a fifteen-meter tsunami flooded the complex, knocking out its power supply. With the system used to cool the fuel rods inoperable, three of the reactors went into meltdown. The reactions caused the emission of hydrogen, and between March 12 and March 15, explosions tore the roofs off the plants' buildings.

The blasts and subsequent steam produced by attempts to cool the exposed reactors with seawater contaminated hundreds of square kilometers with radionuclides. Towns around the reactors were evacuated and placed off-limits. In Tokyo's drinking water, radioactive iodine was detected at double the safe levels for babies; dangerous radiation readings were found in mushrooms, spinach, rice, and seafood. The Pacific was contaminated with massive volumes as five hundred tons of groundwater a day flowed through the basements of the damaged reactors and then ran untreated into the sea.[39]

## 3.11 AND OPERATION TOMODACHI

Following the tsunami, the US military dispatched relief teams to Tohoku in a mission called "Operation Tomodachi." It was brave, essential work that helped thousands of Japanese survivors—but at the same time it showed that the military had learned little about how to deal with radioactive contamination six decades after the debacle of Operation CROSSROADS.

During Tohoku relief operations, service members and vehicles became contaminated with radiation that was then spread far and wide throughout its bases in the mainland and Okinawa. According to briefings from the military's Joint Task Force Civil Support (JTF-CS), a group specializing in chemical, biological, and radiological incidents, among the vehicles contaminated in Japan were Humvee trucks and Sea Knight helicopters. Radiation accumulated in the vehicles' air filters and wheel wells—and also on the blades of helicopters. The report stated that some equipment was "permanently damaged beyond repair."[40]

Another report from the US Army's Center for Army Lessons Learned (CALL) shows the primitive methods by which the military attempted to decontaminate its vehicles. Service members used paper towels, "baby wipes," duct tape, and hot water; it seems some were not equipped with protective clothing. The reason given for such basic techniques was that they were "based upon an effort not to alarm the civilian population." The contaminated water and solid waste were stored on six installations: Misawa Air Base, Yokota Air Base, NAFA, Yokosuka Naval Base, Sasebo Naval Base, and MCAS Futenma.[41]

One internal document revealed the large volumes involved. As of May 2011, Misawa Air Base had more than 30,000 liters of liquid waste, and Atsugi Air Base had almost 95,000 liters of liquid waste and 37 barrels of solid waste; there were smaller volumes, too, at Yokota Air Base and Sasebo Naval Base. In response to my enquiries, the military admitted that radioactive wastewater had been dumped at the Misawa and Atsugi bases; it claimed the liquid was "determined to be safe based on GOJ standards for discharge into the sewer system." Neither of the communities near the bases had been notified of the disposal, and when my investigation made headline news, there was widespread anger.[42]

Many US service members believe they were sickened during relief work. Operating off the coast of Tohoku, there were twenty-five navy vessels—including the largest, the USS *Ronald Reagan*—with approximately seventeen thousand crew; when explosions tore apart the Fukushima power plant, radioactive materials drifted out to sea. One of those on board the Reagan recalled the clouds as a warm gust in the freezing cold air; it felt "like I was licking aluminum foil," she said.[43]

The ship's commanding officers announced they had entered a radioactive plume and ordered everyone below decks. Ventilation systems were turned off, but, much to the crew's dismay, they were soon ordered back into the area where they had encountered the cloud. Other ships arrived in the vicinity as well. Just like those involved in Operation CROSSROADS, their vessels' drinking water was produced from desalinated seawater, and this too had become contaminated with radiation. The crews drank it, ate food that had been cooked in it, and showered in it.[44]

One sailor recalled what he believes was the immediate effects of exposure, saying, "People would shit themselves on the flight deck so often that it wasn't even a surprise anymore." In addition to diarrhea, other short-term symptoms included hair loss and rashes. As sickness spread throughout the crew, the navy blamed it on stress and then gastroenteritis. With the broken reactors continuing to emit radioactive steam and particles, the navy vessels remained in the area conducting relief operations for the next several weeks; according to sailors, they were sometimes so close to shore that they could see the nuclear power station. Service members were ordered to decontaminate the decks; photos from the work show them scrubbing with brushes; no one is wearing masks or ventilators.

In 2014, the assistant secretary of defense for health affairs, Dr. Jonathan Woodson, concluded the exposure of Operation Tomodachi participants was "very small and well below levels associated with adverse medical conditions."[45] In the following years, sailors fell ill from thyroid disorders, leukemia, and other cancers; at least seven died. Unable to sue the Department of Defense, they filed lawsuits against TEPCO at a court in San Diego, where, at the time, the USS *Ronald Reagan* had been homeported. The number of sailors involved in the case reached four hundred; however,

On March 22, 2011, sailors aboard the USS Ronald Reagan (CVN 76) scrub decks to remove radioactive contamination following the triple meltdowns at the Fukushima Daiichi power plant; hundreds believe they were sickened by exposure. US Navy photo by Mass Communication Specialist Seaman Nicholas A. Groesch

in March 2019, the court dismissed the case, with the judge stating, "After considering the Japanese and United States governments' views, the Court finds that the foreign and public policy interests weigh toward dismissal."[46] The handling of the case has infuriated veterans and their legal team; one of their lawyers stated, "These kids were first responders. They went in happily doing a humanitarian mission, and they came out cooked."[47]

Although the US military, like the Japanese government, often tries to downplay the effects of TEPCO's radiation, its own data contradict these reassurances. Five years after the 3.11 meltdowns, sixteen of the twenty-five US ships that participated in relief operations remained contaminated, notably the USS *Ronald Reagan*, which despite undergoing two extensive cleanups was still suffering from low-level contamination as recently as 2016. According to USFJ, the volume of radioactive waste at Yokosuka Naval Base is continuing to accumulate as ships that participated in Operation Tomodachi undergo maintenance at the installation.[48]

Given its vested interests in the nuclear power industry, the Japanese government has repeatedly downplayed the impact of the meltdowns at Fukushima. In 2013, Prime Minister Abe Shinzo stood in front of the International Olympic Committee and assured them that the situation was under control. It was a bold-faced lie that made him the laughingstock of many Japanese.

As of 2019, approximately six hundred tons of melted fuel remain at the bottom of the Fukushima reactors, emitting radiation strong enough to paralyze robots. In 2017, TEPCO announced readings of 530 sieverts (Sv) an hour; to put this in context, experts estimate human exposure to just one sievert can cause hair loss, cataracts, and infertility, and four sieverts would kill 50 percent of those exposed. None of this fuel has been removed—and the technology does not yet exist to begin operations.[49]

Groundwater continues to flood the reactors' basements at the rate of approximately one hundred tons per day, contaminating it with radionuclides; water pumped into the reactors to cool the fuel becomes contaminated, too. In 2019, TEPCO had 1 million liters of wastewater stored at the plant in hundreds of plastic tanks. TEPCO claimed it had treated the water with a filter system to remove all but one radionuclide—tritium—but in 2018, the water was still found to contain dangerous levels of iodine, cesium, and strontium. As recently as February 2019, levels of cesium exceeding safe limits were discovered during test fishing off the coast of Fukushima. The Japanese government is planning to release the contaminated water into the Pacific Ocean.[50]

On land, too, despite government assurances that decontamination efforts were progressing smoothly, hot spots remain scattered throughout east Japan, including in Tokyo river sediments, and attempts to clean up forests in Tohoku have not even started in many places, resulting in radionuclides flowing down mountainsides into rivers and communities after heavy rains. Workers involved in decontamination are exposed to hazardous levels of radiation and often exploited by subcontractors, some of whom are affiliated with the Yakuza. Millions of tons of contaminated soil sit outside in plastic sacks in Fukushima, and at the time of this writing, the Japanese government is pushing through plans to reuse it in construction projects throughout the nation, spreading radiation far and wide.[51]

Despite these ongoing problems, the Nuclear Village—and especially the Japanese government—has continuously pressed for reliance on nuclear power. In a series of failed export plans, it attempted to sell reactors to Turkey, Vietnam, and the United Kingdom. In Japan, it has pushed forward with the restart of reactors that had been closed following the March 2011 meltdowns. In the contaminated communities in Fukushima, the government has declared many cleanup operations complete and lifted restrictions on the return of some evacuees. To facilitate these returns, the Japanese government raised its safe level from 1 millisievert (mSv)/year to 20 mSv/year; as with many other forms of contamination, children are particularly susceptible to radionuclide hot spots—their bodies are still developing and they often play outdoors in polluted sand and soil. UN human rights special rapporteur Baskut Tuncak has criticized the Japanese government's moves as "deeply troubling, highlighting, in particular, the potentially grave impact of excessive radiation on the health and well-being of children."[52]

## LSD, PLATE TECTONICS, AND MILITARY NUCLEAR RISKS

The scale of the tsunami's damage in March 2011 awoke many Japanese people to the vulnerability of its Pacific coastline and the dangers of locating nuclear reactors there. In the same way that the exposure of the *Lucky Dragon* united the public against nuclear weapons, the March 2011 meltdowns caused large-scale opposition to nuclear power. In Tokyo, tens of thousands of people took to the streets to oppose LDP government plans to restart nuclear power stations. At the same time, it reignited concerns about the safety of anchoring US nuclear-powered vessels at Japanese ports. Nuclear submarines frequently visit the bases, and, since 2008, Yokosuka

Naval Base has been the only overseas homeport for US nuclear-powered aircraft carriers: first, the USS *George Washington* between 2008 and 2015, and then the USS *Ronald Reagan*.[53]

At more than three hundred meters long and twenty stories high, these Nimitz-class ships hold two nuclear reactors fueled by highly enriched uranium. Whereas most civilian reactors run on uranium enriched to between 3.5 and 5 percent, these are enriched to 93 to 97.3 percent—the same concentration from which nuclear weapons are made. Such fuel can ignite if it comes into contact with air. The exact weight of the uranium in the twin reactors is classified, but its power—and danger—can easily be envisaged from available information. The ships can operate for twenty years and sail tens of thousands of kilometers on only one fueling. The energy the reactors produce is enough to power a small city.

Even before the deployment of these ships to Yokosuka Naval Base, the military was trying to convince the Japanese people of the safety record of its nuclear-powered fleet. In December 2005, they boasted to the *Asahi Shimbun*, "A radiation related emergency is almost impossible."[54] The Japanese government has never questioned such assurances. In April 2006, it uploaded a document titled "Fact Sheet on US Nuclear-Powered Warship (NPW) Safety" on the Ministry of Foreign Affairs homepage, which remains its current stance. Basing its advice on a 1967 US government memo, the paper promised, in the case of a nuclear accident aboard a vessel,

> The maximum possible effect of the predicted amount of radioactivity released would be localized and not severe: The effect would be so small that the area where protective actions, such as sheltering, would be considered at all would be very limited, and only in the immediate vicinity of the ship and well within the US Navy bases in Japan.[55]

Suspicions that the Japanese government does not take the risks of such vessels seriously surfaced again in May 2014, when it was revealed the radiation levels for evacuation from vessel-related accidents had been set at 100 microseiverts (µSv)/h—far higher than the 5 µSv/h level for civilian power stations.[56] Such a position is criminally negligent. It swallows whole the Pentagon's long-perpetuated myth of the safety of its vessels and, even more naively, encapsulates how the Japanese government assumes the wire fences encircling military installations can keep contamination from harming neighboring communities. In particular, the danger of stationing nuclear-powered vessels at Yokosuka can be reduced to two factors: military incompetence and plate tectonics.

As previous chapters have shown, military operations have been riddled with carelessness; the US Navy is no exception. Since the 1990s, nuclear-powered vessels have been involved in several major accidents caused by human error. In February 2001, the *Ehime Maru*, a fishery high-school training ship, was sailing off the coast of Hawaii when a nuclear-powered submarine, the USS *Greeneville*, suddenly surfaced directly beneath her hull. The Japanese ship sank, and nine of those on board—four of them children—died.

Among the reasons for the accident was the distracting presence of civilian VIPs aboard the submarine, two of whom were allowed to operate the system that surfaced the submarine beneath the *Ehime Maru*. The submarine's commanding officer only received a minor punishment. The VIP tour itself had been arranged by retired admiral Richard Macke, the commander who had lost his previous job for his comments following the 1995 gang rape on Okinawa.[57]

In May 2008, just prior to her scheduled arrival at Yokosuka Naval Base, a serious accident occurred aboard the USS *George Washington*—again due to negligence. Sailors smoking near improperly stored flammable chemicals sparked a fire that raged for 8 hours, burned through 8 decks, and injured 37 sailors; fortunately, it did not reach the two reactors. The fire was one of the worst in the US Navy's peacetime history and cost $70 million in repairs.

In 2017 alone, there were three major accidents involving US vessels homeported at Yokosuka, resulting in the deaths of seventeen US sailors and the dismissal of the commander of the Seventh Fleet. These ships were fitted with state-of-the-art navigation equipment that ought to have been foolproof—but they still crashed.[58]

Perhaps most disturbingly, in 2018, the Naval Criminal Investigative Service uncovered an LSD drug ring involving more than a dozen sailors assigned to the USS *Ronald Reagan*'s nuclear reactor division. The potential risks of mixing psychedelics and enriched uranium in confined spaces are unfathomable.[59]

The area in which the Seventh Fleet operates includes Tokyo Bay, one of the most congested seaways in the world. Were a nuclear aircraft carrier to collide with one of the many oil or liquefied petroleum gas (LPG) tankers that traverse this area each day, the result would be apocalyptic. Further reason to question the safety of docking nuclear-powered vessels at Yokosuka comes from beneath the ground. Numerous fault lines run around the base. In 1703, a magnitude 8.0 earthquake—known as the Genroku Great Earthquake—struck the area, killing thousands of people. The earthquake triggered a tsunami ten meters tall and damaged four hundred kilome-

*Oil booms deployed around a spill from the USS Fitzgerald in 2011; identities of two sailors on deck redacted by FOIA officer for privacy reasons. In 2017, seven sailors were killed aboard the vessel when it collided with a container ship.* Obtained via the Freedom of Information Act

ters of coastline. The reoccurrence of a similar scale quake could cause a tsunami strong enough to damage a nuclear-powered vessel. A large wave could incapacitate or even run aground any military submarines operating close to the harbor. Moreover, at the berth for Nimitz-class aircraft carriers on Yokosuka Naval Base, between the bottom of the ship and the bay, there is relatively little room. A tsunami undertow could leave the ship high and dry, potentially disabling its reactors' cooling systems.[60]

Prior to 3.11, such speculation may have seemed the stuff of a disaster movie; however, the magnitude 9.0 Tohoku quake and nuclear meltdowns taught us the real risks of nature and the folly of human hubris. Even a small fire aboard a navy nuclear reactor could send plumes of highly enriched uranium—far more toxic than that of Fukushima Daiichi—across Tokyo Bay. The consequences for the tens of millions of residents in its path would be catastrophic, and neither the Japanese nor American governments have done any concrete planning to deal with such an eventuality—preferring instead to believe their own myths of safety. Given these risks, it is no exaggeration to say that the stationing of nuclear-powered vessels—aircraft carriers and submarines—at the mouth of Tokyo Bay is a nuclear disaster waiting to happen.

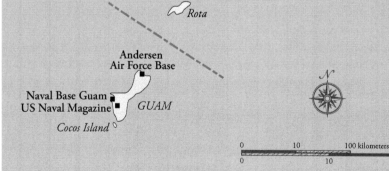

PACIFIC
OCEAN

⟨Pagan

○ Alamagan

○ Guguan

Philippine
Sea

○ Sarigan

Anathan ⟨⟩

○ Farallon de
Medinilla

COMMONWEALTH OF
THE NORTHERN MARIANA
ISLANDS

⟨ Saipan

⟨ Tinian
○ Aguijan

⟨ Rota

Andersen
Air Force Base

Naval Base Guam
US Naval Magazine ■ GUAM

Cocos Island ○

N

| 0 | 10 | 100 kilometers |
| 0 | 10 | 100 miles |

*Guam and the Commonwealth of the Northern Mariana Islands*

# 8

# TOXIC TERRITORIES

## Guam, the Commonwealth of the Northern Mariana Islands, and Johnston Atoll

**F**or many Americans, the word *colony* is anathema. The United States was born out of its revolution against imperialist Britain, and the constitution enshrines the rights to independence, equality, and liberty for all. Few Americans are willing to admit that today, the United States possesses an empire—and perhaps even fewer could list its overseas possessions or how it came to own them.

Following the defeat of Spain in 1898, the United States took over its colonies—the Philippines, Puerto Rico, and Guam—and then a series of Supreme Court rulings known as the Insular Cases established the basis for how these islands would be governed. The judges decided the constitution did not fully apply there, and because they were "inhabited by alien races," ruling "according to Anglo-Saxon principles may for a time be impossible." Upon these racist foundations, the United States founded its overseas empire, whereby such residents would be treated as second-class citizens. Accompanying this decision, in the early twentieth century, the United States adopted a semantic sleight of hand to conceal its imperialism—instead of referring to its overseas possessions as *colonies*, it would use a word free from such negative connotations: *territories*.[1]

Today, the United States owns sixteen territories, five of which are inhabited by a combined population of approximately 4 million people. These residents exist in geopolitical limbo, bearing American passports but lacking full constitutional rights. They can elect a member of Congress, but that delegate cannot vote, only make suggestions. Residents of territories have no representative in the Senate and no right to vote in presidential elections, even though, in Guam's case, they serve and die in the US military at the highest rate per capita.[2]

Territories' lack of democratic rights allows the federal government to operate without accountability to residents, a particularly serious abuse when it comes to Department of Defense operations. In Puerto Rico, the military used Vieques island as a bombing range, contaminating it with dioxin, depleted uranium, and lead; cancer levels there are the highest in the territory, and other serious illnesses are rife. In the Pacific, the Pentagon has taken advantage of geographical isolation and lack of oversight to pollute this region's territories with zero regard for the environment or human health. The Marshall Islands, a US-administered territory until the late 1970s were pummeled by nuclear detonations and fallout for twelve years; Wake Island, still a territory today, was used as a dump for polychlorinated biphenyls (PCBs) brought from US bases in Japan when they had been barred from import to North America, and it continues to serve as a missile test site. Three other Pacific territories—Guam, the Commonwealth of the Northern Mariana Islands (CNMI), and Johnston Atoll—have been impacted by military contamination, but the US government has either been slow to react or not acted at all. Johnston Atoll has suffered the worst; poisoned with plutonium, Agent Orange, and chemical weapons, today it sits as an abandoned sacrifice zone in the North Pacific, a testimony to military hubris.[3]

## GUAM

The Mariana islands—today politically divided into Guam and CNMI—were first settled by the Chamorro people in approximately 2000 BC. A matriarchal society, they lived off farming, fishing, and interisland trade. In 1521, the first European explorer, Ferdinand Magellan, arrived in the Marianas, followed some forty years later by a delegation that declared the islands a colony for Spain, the first of four foreign nations that would lay claim to the region. In the coming years, Chamorros suffered a fate familiar to most indigenous peoples after European encounters—massacres, epidemics, and forced relocation to villages where they could be more tightly controlled; from 200,000 Chamorros at the start of Spanish rule, the population dropped to 5,000.

After the Spanish–American War of 1898, the United States took control of most of Spain's colonies: Cuba, Puerto Rico, the Philippines, and Guam. Spain sold its other Micronesia possessions—the Northern Marianas, the Marshall Islands, and the Carolinas—to Germany. This set the region on two divergent trajectories in the early twentieth century that would briefly merge in the horror of World War II before separating again.

On Guam, following the US takeover from Spain, residents expected to receive American constitutional rights, but instead martial law was imposed on the island under the administration of the US Navy, which ran it with military discipline, punishing petty infractions and banning the Chamorro language in government offices and schools. Residents' repeated appeals for citizenship and the ability to vote for president were denied as a potential threat to military dominance. In 1936, the US Navy cited the "racial problems of that locality" to assert that "these people have not yet reached a state of development commensurate with the personal independence, obligations, and responsibilities of United States citizenship."[4]

On December 8, 1941, just hours after attacking Pearl Harbor, the Japanese military struck—and soon overran—Guam. Renaming it *Omiya Jima*, "the Great Shrine Island," the Japanese military ruled with brutality, locking islanders into prison camps, forcing them to build airfields, and slaughtering them in droves. More than eleven hundred islanders died during the Japanese occupation.

On July 21, 1944, the United States began a 13-day bombardment of Guam, followed by a land battle, which, by its end on August 10, had killed 18,000 Japanese and 1,700 US troops. The Japanese military had so strongly indoctrinated its forces not to surrender that the final Japanese soldier, hidden in the island's jungles, did not give himself up until 1972.

Following the liberation of Guam, residents felt immense gratitude toward US forces—tinged with apprehension, as the military occupied most of the island and developed it into a launchpad for the invasion of Japan. Just as on Okinawa, the military housed residents in internment camps while it seized their land, including some of Guam's best arable and fishing areas, to build bases. In the north, the military constructed North Field, later called Andersen Air Force Base, atop the Northern Guam Lens Aquifer, the island's main source of fresh water. In the south, the military took over the island's only deep-water harbor at Apra, building Naval Supply Depot, a refueling facility that, by July 1, 1945, had dispensed more than 1.5 billion liters of oil and fuel for the war effort. At the same time, electricity was largely restricted to military use, a situation that continued into the mid-1950s.[5]

After the August 1945 surrender of Japan, the military retained approximately 85 percent of the land it had occupied on Guam and began to consolidate its infrastructure, for example, relocating residents of Sumay Village. Apra Harbor was strengthened, and Naval Supply Depot stored materiel for both the US Navy and Air Force, while the US Naval Magazine Guam stretched more than sixty-two hundred acres, the "westernmost ammunition supply point on US soil."

Congressman F. Edward Hebert, who later became the head of the House Armed Services Committee, justified the military's dominance in the region, stating, "We fought for them, we've got them, we should keep them."[6] In many ways, postwar Guam resembled Okinawa. One journalist described Guam as a "vast junk yard and a one-time battlefield where the scars of combat still offend the eye everywhere."[7] The military dumped hundreds of tons of munitions in the seas throughout the region, while many residents of Guam and other islands relied on collecting scrap metal to earn money; following the war, its sale was the region's second largest export.[8]

In addition to contamination from unexploded ordnance (UXO), lead, fuel spills, and solvents, residents had to contend with another hazard brought about by the military: brown tree snakes. Accidentally imported aboard boats and flights from the South Pacific, the snakes proliferated on Guam, today numbering more than 1 million. They have wiped out ten of the island's twelve native bird species, causing an inverse explosion in the spider population, plus deforestation, as trees, once sown by seed-dispersing birds, can no longer spread.[9]

## The Cold War's Hot Wars, Contamination, and Nuclear Weapons

In 1950, Congress passed the Organic Act of Guam, making the island an unincorporated territory; however, the military continued to dominate the island, owning approximately half of its land, and the act allowed the United States to use its powers of eminent domain to legally own the properties taken for the Department of Defense. These bases experienced heavy use during the Cold War. Between 1965 and 1973, B-52 bombers from Andersen Air Force Base flew tens of thousands of missions over Southeast Asia. At its peak in 1972, there were 155 B-52s at Andersen Air Force Base, the highest number in history. These aircraft required constant maintenance, including 120 jet engine overhauls a month; the volumes of oils, solvents, and fuels is incalculable, and these substances still contaminate Guam today.[10]

*A B-52 takes off from Guam's Andersen Air Force Base during Operation Linebacker II, December 1972. At the time there were 155 of the aircraft stationed on the island, creating extensive contamination from fuel, oils, and solvents.* Photo courtesy of US Air Force

While the US government claimed its wars in Korea and Vietnam were fought to secure human rights and democracy, on Guam, the rights of residents were trampled beneath military priorities. A naval security clearance was imposed on the region until 1962, restricting civilian visits, including shipping in Apra Harbor; these controls hobbled imports, exports, and economic growth.

In 1961, the United Nations (UN) criticized the United States for failing to compensate residents whose land the military had seized. When Peace Corps legal teams attempted to teach islanders about property rights, they were criticized by the military, and their work was subsequently phased out by President Richard Nixon.[11]

Beginning with the August 1945 launch of atomic bombers from Tinian island—today part of CNMI—US nuclear operations have dominated the Micronesia region. For much of the Cold War, Guam's Andersen Air Force Base was a Strategic Air Command installation, a key element for waging nuclear war. "Guam's strategic location was an ideal potential launching site for atomic weapons under military operations in the Far East," according to the Department of Defense's official history of the region.[12]

In 1951, the military first brought nuclear weapons to Guam for possible use in Korea; in the following two decades, at least twelve different nuclear weapons were stored on the island, for example, Regulus cruise missiles armed with two-megaton warheads that could be launched from ships or submarines. Between 1964 and 1981, Apra Harbor was the homeport for attack submarines armed with Polaris nuclear missiles. The military also designed an extensive network of nuclear shelters for its personnel on Guam, including two large bunkers built in expanded Japanese World War II shelters on the navy base.[13]

US military operations have contaminated Guam with radiation in numerous ways. Following the 1946 CROSSROADS Able and Baker tests, approximately eighteen vessels from the guinea pig fleet were brought to Guam for attempted decontamination. The work was experimental, unsuccessful, and contaminated the environment; at least one of the vessels was sold for scrap on Guam, potentially spreading radiation to its new owners. Recognizing the dangers of such work, today, the military personnel who took part in decontamination attempts on Guam are eligible for VA support—but no such help is available for civilian workers or local communities.[14]

In November 1952, the United States detonated its first hydrogen bomb, the building-sized IVY Mike, which exploded with a yield of 10.4 megatons, blowing 80 million tons of soil into the atmosphere. On Guam, nineteen hundred kilometers away, US Navy lieutenant Charles Bert Schreiber, an atomic, biological, and chemical warfare defense officer, recorded significant radiation levels on the island following the blast—but when he reported his findings to the military, he was ordered to remain quiet.[15]

As a result, the contamination of Guam remained largely unknown for decades; however, following campaigns by veterans and residents, in 2005, a report by the National Research Council concluded, "Guam did receive measurable fallout from atmospheric testing of nuclear weapons in the Pacific. Residents of Guam during that period should be eligible for compensation under RECA (Radiation Exposure Compensation Act) in a way similar to that of persons considered to be downwinders."[16] Supporting the decision was a blue-ribbon panel formed by Congress in 2010, which stated the military had "put the population of Guam in harm's way knowingly and with total disregard for their well-being."[17] Despite these findings, as of 2019, residents of Guam had still not been awarded help for their exposure.[18]

Radioactive contamination from nuclear-powered vessels has also leaked into Guam's sea. In 1975, coolant water spilled from the submarine tender USS *Proteus* into Apra Harbor. The leak reportedly caused Geiger counters to spike at 100 millirems per hour, fifty times the safe limits. The

USS *Houston*, the submarine that contaminated Okinawa and Japan, was homeported at Apra Harbor. Between 2006 and 2008, it was on Guam for 366 days, during which it leaked radiation. In 2014, the USS *Jefferson City* was stranded in Guam for five months following a leak of coolant water in its nuclear propulsion system. The military admitted the coolant contained "trace amounts of radioactivity" but said there was no risk to human health or the environment.[19]

Today, the United States Air Force (USAF) maintains what it calls a continuous bomber presence on Guam, consisting of B-52s, B-1s, and B-2s, all of which are capable of carrying nuclear weapons. Their presence is no secret. On Guam's Channel 32 military TV station, one USAF TV spot showed nuclear devices on several air bases with the following commentary:

> We serve in the middle of nowhere on the edge of history. . . . We are not out here babysitting metal, twiddling thumbs, or flying in circles. . . . We are here to scare the living hell out of our enemies. Scare them with our power, our resolve, our dedication, our vigilance. We will not be the ones to blink.[20]

Sometimes, however, their presence threatens those on the ground.

In February 2008, a B-2 crashed during takeoff from Andersen Air Force Base. The two pilots safely ejected, and the USAF announced the aircraft had not been carrying nuclear weapons at the time. A subsequent investigation blamed the crash on moisture in the aircraft's sensors, and it has been dubbed the most expensive aviation accident in the world: an estimated $1.2 billion. Two years later, another B-2 on Guam experienced a major fire, but the USAF downplayed its severity to the public; the truth about the incident, which almost destroyed the aircraft, only emerged the following year.[21]

## AGENT ORANGE AND MILITARY HERBICIDES ON GUAM

According to the Department of Defense, large volumes of defoliants were brought to Guam. "In 1952, roughly five thousand drums of Herbicide Purple were transported to Guam and stored there in anticipation of use on the Korean Peninsula." the undersecretary of defense wrote to Senator Lane Evans in 2003.[22] Consisting of the two herbicides 2,4-dichlorophenoxyacetic acid (2,4,-D) and 2,4,5-trichlorophenoxyacetic acid (2,4,5-T), Agent Purple was a forerunner of Agent Orange. Approximately 2 million liters were sprayed over Vietnam from 1962 to 1965, and experts believe it was more contaminated with dioxin than defoliants produced in later years.

The Department of Defense claims Guam's Herbicide Purple stockpile was never used and sent back to the United States; it had also stated it has no records of other defoliants on Guam. However, service members stationed there during the 1960s and 1970s claim large stockpiles of defoliants, including Agent Orange, were stored on Guam, particularly at Andersen Air Force Base. These veterans say they sprayed defoliants along the fuel pipeline that ran from Apra Harbor to the air base, and they also sprayed the runways' edges to keep them clear of vegetation.[23]

Edward Jackson, a sergeant with the 43rd Transportation Squadron, assigned to Guam in the early 1970s, recalled, "Andersen Air Force Base had a huge stockpile of Agent Orange and other herbicides. There were many, many thousands of drums. I used to make trips with them to the navy base for shipment by sea." Jackson remembered dumping military waste, including damaged barrels of defoliants, over cliffs.[24]

Leroy Foster, a master sergeant in the USAF, described the "vegetation control duties" he performed following his assignment to Andersen Air Force Base in 1968. "I mixed diesel fuel with Agent Orange, then I sprayed it by truck all over the base to kill the jungle overgrowth," he said. "None of the older service members wanted to do the work, so because I was the low man on the totem pole, it was left to me." Soon after starting this work, Foster suffered serious skin complaints, and in the following years, he fell sick with Parkinson's and ischemic heart disease. His daughter developed cancer as a teenager, and his grandchild was born with twelve fingers, twelve toes, and a heart murmur.[25]

In 2015, researchers, including those from the Guam Department of Public Health and Social Services, discovered the communities where defoliants may have been sprayed had suffered high incidences of infant deaths from birth defects.[26] Although US authorities have repeatedly denied defoliants were sprayed on Guam, the VA has awarded help to at least twelve veterans exposed on the island, one of whom was Foster. He died in 2018.[27]

The similarities to the use of defoliants on Okinawa are appalling—service members sickened following orders but today denied help by their own government. Nonetheless, because Guam is a US territory, the government has had to make perfunctory efforts to respond to public concerns. In 2017, US authorities launched an investigation into US defoliant use on Guam, but it has come under fire. For example, critics grew suspicious when the Department of Defense announced that tests on soil did not contain herbicides but Environmental Protection Agency checks revealed their presence.

Guam senators have backed bills in Congress for the territory's inclusion on the VA's list of places where Agent Orange was used. In March 2019, a bill named after Lonnie Kilpatrick, a service member sickened on Guam who died in 2018, aimed to extend compensation to fifty-two thousand veterans exposed to defoliants in three US territories—Guam, American Samoa, and Johnston Atoll—between 1962 and 1980.[28]

## The Toxic Tip of the Spear

For decades following the end of World War II, the military on Guam disposed of its waste without any understanding or concern about its harmful effects on the environment, service members, and residents. Many of its worst practices took place atop the island's drinking water aquifer at Andersen Air Force Base.

In 1978, there was the first major revelation about military contamination when the solvent trichloroethylene (TCE) was detected in wells supplying drinking water on the base. One well peaked at 29.9 parts per billion (ppb), whereas the other ten wells were contaminated "on a sporadic basis" above the EPA's guideline of 5 ppb. No tests had been conducted prior to 1978, so service members must have been consuming tainted water for many years.[29]

According to the General Accounting Office (GAO), TCE had likely entered the water system via approximately twenty abandoned dump sites at Andersen Air Force Base or through the disposal of hazardous chemicals into its storm drainage system. Following the discovery of the contamination, the military tried to lower TCE levels by diluting the highly contaminated water with water that was less polluted; however, a miscalculation resulted in more than twelve hundred service members and their dependents receiving water that was still severely contaminated. Furthermore, after the initial discovery of TCE contamination in 1978, the base failed to perform regular checks on the wells, and even when it did so, test containers broke on the way to the laboratory, rendering analysis impossible.

In later years, the military and the EPA conducted surveys on Guam to investigate the extent of contamination from other substances, but the military repeatedly attempted to conceal or downplay the damage. For example, in the 1980s, surveys conducted by the US Navy and USAF did not consult the EPA, and the military failed to check diligently for contamination.

Despite these tactics, by the mid-1990s, a clearer picture had begun to emerge of how military operations were impacting Guam's environment. On the bases, there were 155 suspected contaminated sites, and outside the bases there were forty-seven.[30] In 1992, Andersen Air Force Base was added to the EPA's Superfund list. It had almost fifty contaminants of concern, including fuel, PCBs, pesticides, and heavy metals. For decades, hazardous waste had been tipped over cliffs around the perimeter of the facility onto the land or sea below.[31]

In 1999, approximately thirty-five chemical weapon test kits, containing small volumes of mustard agent and phosgene, were unearthed on private land. The discovery sparked another GAO investigation in which residents expressed their dissatisfaction with the way the military handled off-base contamination. Three main concerns were highlighted: an opaque system of how the military added new sites to its list of those to be cleaned up, failure to remediate nonhazardous waste, and the slow pace of funding cleanups.

Accusations that the military was shirking its financial responsibilities arose again in the 2000s, concerning the cleanup of Ordot dump in central Guam. Before and after World War II, the military had used the area to dispose of waste, and then it passed along the site for civilian use. By 1986, the dump was leaching pollutants into a nearby river, leading the federal government to demand its closure. In 2011, the government of Guam finally shut down the dump, capped it, and opened a new site at a cost of more than $160 million, but the military refused to pay for any of the remediation costs. In March 2017, the Guam attorney general took the unprecedented step of suing the US Navy.[32]

## PCBs

On Guam, the military has caused widespread PCB contamination throughout the island. In 2012, the EPA removed approximately 320 tons of PCB-contaminated soil from a former military pump station at Agana Springs after the Department of Defense refused to take responsibility for the cleanup.[33] Worse PCB problems have impacted the village of Merizo on the southern shore of Guam, where, in 1944, the Coast Guard began operating a long-range navigation station from the island of Islan Dåno', known in English as Cocos Island. In 1962, a typhoon wrecked the facility, strewing debris, including transformers filled with PCB-laden oil, across the land and adjacent lagoon. The station closed in 1965, and for decades, contamination seeped from the abandoned transformers, covering a wide area.[34]

Residents were largely unaware of the problem until 2005, when tests revealed high levels of pollution from PCBs, lead, and cadmium. Moreover, eleven of the twelve most commonly consumed fish also registered PCBs as high as 250 times the recommended levels. In 2006, authorities placed an advisory on Cocos Lagoon, warning people not to eat fish caught there. In 2007, some contaminated soil was removed from the island; however, in the following years, PCB contamination persisted, and, in 2015, a new contaminant was detected: dichlorodiphenyltrichloroethane (DDT). As of 2019, the fishing ban was still in place.[35]

Research conducted by the University of Hawaii suggested Merizo villagers' health may have been impacted by the contamination in Cocos Lagoon. The 2011 report revealed that cancer deaths between 1978 and 1997 were almost double those of other villages on the island. The authors concluded that although their investigation "does not conclusively link PCB contamination of Cocos Lagoon to an increase in cancer mortality among the residents of Merizo, the temporal relationship is intriguing and certainly justified public health concerns and the effort to mitigate this environmental hazard."[36]

## Guam: From Trailer Park to the Center of the Radar Screen

More than twenty-five years after cleanup work began on Andersen Air Force Base, it has still not been completed. As of 2017, there were approximately one hundred contaminated sites, eleven of them off-limits. In 2019, elsewhere on the island, there were twenty-three formerly used defense sites (FUDS) where surveys were taking place to ascertain contamination; almost weekly, UXO continues to be discovered both inside and outside military facilities.[37]

In 2016, per- and polyfluoroalkyl substances (PFAS) contamination was discovered in three water wells at levels as high as 410 parts per trillion (ppt)—the EPA advisory is 70 ppt—forcing their closure. Also at sea, military operations are causing concern, as the navy has doubled the size of its training zone. Whales have started to beach on Guam—a phenomenon almost unheard-of in the past—leading experts to suspect the blame might lie with the navy's use of sonar near the island.[38]

Many Guam residents fear that exposure to military contamination during a span of many years has damaged their health. Between 2003 and 2007, rates of mouth, nasopharynx, and liver cancers were higher than in the United States, and the diabetes rate is about five times that of the United States. Worsening the problem is the diet of many residents. With the military dominating the best arable land and shorelines occupied by bases and naval training, disrupting fishing areas, people are almost entirely dependent on imported, often processed, food.[39]

In the coming years, the military is set to bring more environmental problems to the island. In 2004, the commander of the naval station on Guam, Rear Admiral Arthur J. Johnson, said, "Guam is no longer the trailer park of the Pacific. Guam has emerged from backwater status to the center of the radar screen."[40] Johnson was referring to the military's buildup on Guam and the Mariana region, one of the largest shifts ever attempted by the Department of Defense, involving the expansion of Andersen Air Force Base and US Naval Base, plus the relocation of eight thousand Marines from Okinawa.

In preparation for the buildup, in 2009 the military released an 11,000-page Draft Environmental Impact Statement and gave the public just ninety days to comment on it. For the first time, the report revealed the true extent of the Pentagon's plans: A live-fire range would be built on the ancient village of Pagat; 22 more wells would be sunk into the aquifer; a further 2,200 acres would be acquired for operations; and there would be an 80,000-person increase in the island's population, a rise of almost 50 percent.[41]

Chamorro anger was unprecedented. Ten thousand public comments were lodged criticizing the plans, and the EPA slammed the report as "environmentally unsatisfactory" and gave it the worse grade possible. The agency was particularly concerned by the military's failure to take into consideration the impact on water supplies and the need for wastewater treatment. The National Trust for Historic Preservation and the civic group We Are Guahan sued the Department of Defense concerning its plans to train at Pagat.

As a result of the outrage, the Department of Defense was forced to reassess the buildup. It decided to move forty-eight hundred marines to Guam, with others sent to Australia and Hawaii. The plan to destroy Pagat was also dropped, but in its place a new range would be built on Ritidian Point, a pristine beach and jungle. The shooting range will fire 6.7 million shells a year, threatening the aquifer below with contamination from lead and other heavy metals. Likewise, the core of the buildup remained unchanged, albeit spread out over a longer time frame; the first marines are expected to arrive at their new base in the mid-2020s. The military is keen to emphasize its attempts to mitigate future environmental damage, but residents remember such broken promises in the past, and visiting Okinawan civic groups have made them aware of United States Marine Corps (USMC) abuses at Tori Shima, Marine Corps Air Station (MCAS) Futenma, and Camp Kinser.

In an interview in 2019, Senator Therese Terlaje encapsulated the opinion of many Guam residents when she said, "If we can't trust the DoD to tell the truth about past activities, then it's difficult for us to trust them now and in the future."[42]

## CONTAMINATED COMMONWEALTH: THE NORTHERN MARIANA ISLANDS

Following the 1898 Spanish–American War, Spain sold the Northern Marianas, the Marshall Islands, and the Carolinas to Germany. Until World War I, Germany governed the fourteen Mariana Islands; however, their control of the islands was short and left little lasting impression. In October 1914, Japanese forces landed on Saipan, the third foreign power to colonize the island. During its thirty-year occupation, Japanese authorities attempted to subjugate the region into its empire—Japanese was taught in schools, and tens of thousands of civilians, many from impoverished Okinawa, moved to the islands to grow sugarcane. The League of Nations mandate that granted Japan control of the islands after World War I prohibited Japan from building military installations there, but by the 1930s, Japan had constructed numerous facilities throughout the region, and from there attacks were launched on Pearl Harbor, Guam, and the Philippines.

On June 13, 1944, the US military began its assault on Saipan, firing some 8,500 tons of ammunition at the island; Japan unsuccessfully dispatched scientists to the island to spread fleas infected with the plague. US marines landed on June 15, and in the following three weeks of combat, more than 3,000 American and 30,000 Japanese troops died. Approximately 20,000 civilians also lost their lives, many of whom committed suicide by jumping from cliffs to avoid capture by Allied forces, who they had been told would rape and torture them.

Located about ten kilometers southwest of Saipan, Tinian Island had been cultivated by the Japanese for sugarcane production, and, by 1944, it was home to approximately 15,000 civilians and 8,000 Japanese troops. Between July 24 and August 1, it was attacked by marines, resulting in the loss of more than 320 US service members, 5,500 Japanese military personnel, and approximately 4,000 civilians.

Following the battle, the United States converted Tinian into a base, crisscrossing it with roads and runways. Due to its shape and size similar to Manhattan, they named its thoroughfares after New York's roads—for example, Broadway and Eighth Avenue—names that remain today. Within months, the island had 150,000 US troops and hundreds of B-29 bombers, making it the busiest airport in the world; in August 1945, two of these aircraft carried atomic bombs to Japan. The loading pits of Little Boy and Fat Man still stand on the island, preserved as national historic monuments, framed in glass and surrounded by visitors' origami cranes.

World War II left its mark on Tinian in other ways. One postwar survey identified twenty-three potentially contaminated areas on the island, notably an ocean dump site containing sunken vehicles and an aircraft junkyard called the "West Field Boneyard." In addition there are the residues from the DDT sprayed twice monthly around the camps; UXO; and contamination from sodium arsenite, a chemical used to disinfect pit latrines, recently identified as an "island-wide hazard both in soil and groundwater."[43]

On Saipan, combat caused persistent environmental problems, too. In 2016, an investigation conducted by the University of Guam detailed how detritus, including UXO and other metallic waste, was bulldozed into the sea, buried in caves, or burned following the war. Researchers identified twenty-four locations on land and two sites at sea where there was contamination from heavy metals. At one of the sites, Agingan Point, levels of lead were "alarmingly high," and it had seeped into the nearby lagoon; researchers also expressed concern that fish, crabs, and fruit might have been affected. They concluded "several sites on Saipan could present a significant risk to local residents and wildlife, with a number of soil- and sediment-associated metals occurring at levels that are currently considered unacceptable."[44]

## The CIA and Contamination on Saipan

Following World War II, the Mariana Islands—like the Marshall Islands—were designated members of the Trust Territory of the Pacific Islands, a UN-administered body. But the United States maintained real control under the Security Council, where it wielded a veto. In 1950, with China fallen to Mao Zedong and the Korean War raging, the CIA decided to construct a base on Saipan to train anti-Communist guerillas. To ensure secrecy, in 1952 the island was put under navy jurisdiction, and throughout the region, Naval security clearance was maintained to limit access.

Under the guise of the Naval Technical Training Unit (NTTU), the $28 million CIA base on Saipan was a sprawling complex resembling a "transplanted California suburb," or as one former CIA agent put it, a "miniature Fort Peary," the CIA's training school in Virginia. There were numerous compounds housing trainees from different countries, garages, and fuel depots, as well as a school for Chinese guerillas with more than thirty staff; the NTTU also included Chinese-style buildings designed by a Chinese Nationalist general. The CIA operated its own airstrip at Kagman, on the east of Saipan, and managed a five-hundred-ton ship, the *Four Winds*.[45]

The NTTU was the CIA's most important base in a region that also included outposts in Taiwan, Camp Chinen on Okinawa, and Atsugi-Chigasaki on mainland Japan. On Saipan, hundreds of Chinese guerillas received training in explosives, small arms, parachutes, and radios, which the CIA hoped would enable them to spark a popular uprising in their homeland; however, such use of Saipan violated UN agreements, so when inspectors visited the island, the CIA had to conceal its activities, only re-starting them after the inspectors had departed.[46]

The CIA's hopes that their trainees would be able to overthrow Mao were illusory—a combination of their own wishful thinking and, as on Okinawa, exaggerated promises from disgruntled Chinese Nationalists. The guerillas were met with little success; after parachuting into China, most of them disappeared, never to be heard from again.

Gradually, the CIA began to gear down its operations and withdraw from Saipan; the island returned to Department of the Interior control in 1962. During this period, one resident recalled the agency's abandoned vehicles "littered the island. Just find gasoline and oil and a live battery, start one, and you could take it home."[47]

Unfortunately for Saipan's residents, vehicles were not the only things left behind at the NTTU. At Tanapag Fuel Farm, for example, there were more than forty large fuel tanks, slowly corroding, leaking their contents, and eventually crumpling; one resident remembered wading through oil-contaminated shoreline to catch crabs, and another watched fuel spill onto the land as some of the tanks were demolished to create a residential area. The tanks littered the community well into the 2010s, when they were fi-nally removed or rusted beyond recognition.[48]

## PCBs at Tanapag

During the 1960s, the Department of Defense shipped to Saipan approxi-mately two hundred electrical capacitors originating from Kwajalein Atoll. The one-meter-tall, 225-kilogram appliances contained PCB-contaminated oil. Whether the military planned to dump them on Saipan or only tempo-rarily store them is unclear; they ended up stacked near antennas belonging to Voice of America in Tanapag Village.[49]

In the following years, the capacitors became a common site throughout the community. Some were set as boundary markers or windbreakers, while others were used as stands for barbecues or headstones at the cemetery; in some cases, their shiny inner linings were stripped and used as decorations. Wherever the capacitors stood, they leaked carcinogenic oil.

Villagers began to grow suspicious that the capacitors were making them sick, so in 1988, Saipan's environmental division ran tests; soil in the village spiked at 55,000 parts per million (ppm)—the cleanup standard is 1 ppm—and groundwater contamination later measured 9 ppb, 18 times the safe level.

When the military was asked to remediate the contamination, what followed was an all-too-familiar pattern of foot-dragging and shirking of responsibility. At first, the Department of Defense denied the capacitors had belonged to them, and then they merely moved the capacitors without taking away the contaminated soil. The army attempted two experimental cleanups—bioremediation in 1995 and thermal desorption in 1997—both of which failed.

In 1999, the Department of Defense shipped out approximately 450 tons of contaminated soil for disposal on the US mainland and announced the cleanup was complete; however, the military had left large piles of contaminated soil at the village's cemetery. The ground was found to be PCB-polluted at 25,000 ppm—forcing the cemetery's closure near All Saints' Day, a time when the predominantly Catholic islanders visit family graves. Land crabs—a popular food—were found to be contaminated, and blood checks on villagers revealed high levels of PCBs.

In 2003, the army performed a final cleanup—fifteen years after the contamination had first been detected. Overall, forty thousand tons of contaminated soil had been treated at a cost of $20 million.

The cleanup of Tanapag created widespread anger on Saipan; it also increased awareness of military contamination and the need for further investigations. In 1999, twenty-five other military dump sites were identified on the island. In 2000, the EPA announced checks of the groundwater for Agent Orange contamination following suspicions barrels of the defoliant had been buried beneath a golf course. Then, in 2002, contamination was discovered at I-Denni, a site where the military had dumped and burned waste, resulting in contamination from arsenic, lead, and PCBs. The subsequent cleanup of the site took ten years and $3 million. UXO continues to plague the territory.[50]

## Military Buildup and Live-Fire Ranges

In 1975, the residents of the Northern Marianas voted to become a commonwealth of the United States, and in return for citizenship and federal funding, they signed deals with the military. They leased the island of Farallon de Medinilla as a bombing range and two-thirds of Tinian Island for the construction of a base with the understanding they would be able to access its medical facilities. That installation was never built, leading to criticism that the military had broken its promise; instead, the Department of Defense opted to stage sporadic training on Tinian.

In recent years, the US government announced that CNMI would be built up militarily: Training on Tinian would expand to as many as forty-six weeks a year, the bombing of Farallon de Medinilla would double, and operations would encompass the island of Pagan. When the Department of Defense released its draft environmental impact statement in April 2015, it received 27,000 comments—the total population of the CNMI is only 53,000. The EPA weighed in with its own worries, particularly regarding the fact that live-fire training could potentially contaminate Tinian's aquifer with munitions and rocket fuel. Training might also damage historical sites, for example, the atomic bomb-pits where the Japan-bound weapons were loaded.[51]

Plans to bomb Pagan were especially controversial. The island had once been home to three hundred people, but following a volcanic eruption in 1981, residents were forced to evacuate. Historically, Pagan islanders had existed in this way, temporarily leaving the island following eruptions but returning after things had settled down; however, outsiders were ignorant of this past and wanted to put the island to other uses. Following the 2011 Tohoku tsunami, one Japanese company considered dumping debris on the island, and the Department of Defense wanted to use it for war games, conducting aerial bombardments to the north and landings to the south. To justify its plans, the military claimed the island had no residents, a blatant untruth. Although it did not own land on Pagan, Congress could seize it through eminent domain, and without a voting delegate in the House, the will of CNMI residents could be ignored.

The EPA attacked the military's plans, stating they would damage more than 120 acres of marine habitat. Meanwhile, residents took matters into their own hands by announcing a project to return to agricultural lots on Pagan.

At the forefront to protect the island has been the citizens group Pagan-Watch, cofounded by Peter J. Perez. Summarizing the risks to the region in 2015, he wrote that the CNMI Joint Military Training proposal would surround the region with

> live-fire ranges, in Guam to the south, Tinian in the west, FDM and Pagan to the north, and all around us on and in the ocean. The CJMT (Joint Military Training) will have widespread negative consequences on virtually every aspect of life in the CNMI: health, environment, natural resources, economics,

culture, historic preservation, social justice, infrastructure, public safety, and freedom of movement.[52]

## JOHNSTON ATOLL

Johnston Atoll, consisting of four small islands, the largest of which is Johnston Island, is situated approximately thirteen hundred kilometers southwest of Honolulu. Originally called Kalama Atoll by Hawaiians who claimed it for their kingdom, it lacked a resident population primarily due to an absence of fresh water. In 1796, US whalers landed on its main island—then only forty-six acres—followed by the arrival of British naval captain Charles Johnston, who gave the atoll its English name. In the mid-nineteenth century, US miners arrived to strip the island's guano for sale as fertilizer, and following US annexation of Hawaii in 1898, the atoll fell under US control.

After a brief period as a bird sanctuary, beginning in 1934 the island was commandeered by the US Navy, which built up its infrastructure, particularly a runway. During World War II, dredging expanded Johnston Island, and military aircraft and vessels used the island to refuel, a role that continued during the Korean War. By the 1960s, further expansions increased the number of islands in the atoll to four.

### Nuclear Tests

Between 1958 and 1962, the United States conducted twelve nuclear tests on Johnston Atoll. The first two took place in August 1958, during the Operation HARDTACK series, in which warheads were exploded high in the atmosphere. Prior to the tests, scientists had expressed concerns that the blasts might rip holes in the ozone layer, but the military proceeded with its plans. Present on Johnston Island for Teak, the first test, was former Nazi rocketeer Wernher Von Braun, who had been brought to the United States under Operation Paperclip and was now employed by the rocket program. On August 1, 1958, the rocket designed by Von Braun was launched; a malfunction propelled it directly above Johnston Island, where it detonated at a height of 77 kilometers, with a yield of 3.8 megatons. The blast blinded hundreds of monkeys and rabbits carried in monitoring aircraft and disabled radio communications throughout the region. The test reportedly so unnerved Von Braun that he left Johnston Island before the next test. On August 12, the Orange shot exploded with the same yield as Teak but this time at a lower altitude of 43 kilometers, smack-dab inside the ozone layer.[53]

Four years later, ten tests under the DOMINIC series took place in rapid succession between July and November 1962. The reason for the urgency? The United States was racing to beat a ban on atmospheric testing, set to take effect the following year. One of the tests, named STARFISH Prime, exploded with a yield of 1.4 megatons, four hundred kilometers in the atmosphere on July 8, 1962. The explosion lit up the night sky as bright as day in Honolulu for six minutes, and the electromagnetic pulse set off burglar alarms and disrupted power lines and communications there; the blast also disabled several satellites. Another test—BLUEGILL Triple Prime—burned the retinas of two service members on Johnston Island, and the KINGFISH blast sparked a "yellow-white, luminous circle with intense purple streamers for the first minute"; other visual effects lasted for three hours.[54]

Four of the DOMINIC rocket tests conducted on Johnston Island failed, and three of them seriously polluted Johnston Atoll.[55] On June 19, 1962, during the STARFISH test, a Thor rocket's engines cut out one minute into the launch. Before it could veer off course, the safety officer decided to destroy the missile. At least 250 chunks of plutonium-contaminated debris fell on Johnston Atoll.[56] Five weeks later, on July 25, during the BLUEGILL Prime test, another Thor missile malfunctioned on the launchpad, prompting the safety officer to destroy the warhead by radio command to prevent it from detonating. The missile exploded, causing "extensive alpha contamination." According to the Defense Nuclear Agency, "Burning rocket fuel, flowing through the cable trenches, caused contamination of the trenches and the interior of the revetments and all of the equipment contained in them." Surveys conducted the next day found radioactive contamination as high as 1 million counts per minute compared to normal background levels of about fifty counts.[57]

To clean up the site, civilian contractors tried dousing sand with oil, dumping soil and equipment into the sea, scrubbing concrete with solvents, and painting surfaces with "epoxy or latex paint to fix any remaining contamination." Unsurprisingly, following two weeks of such attempts, the site remained highly contaminated, with radiation levels exceeding one hundred thousand counts per minute. Not only were the BLUEGILL Prime decontamination techniques inadequate, but also dumping contaminated soil and debris into the lagoon returned to haunt those on Johnston Atoll. In the following years, sand was dredged from the lagoon to expand the islands, lacing them with particles of plutonium and other radionuclides.[58] Service members present on the island at the time of the accident subsequently suffered diseases they believe were caused by exposure to radioactive contamination, including non-Hodgkin's lymphoma, thyroid cancer, throat cancer, and reproductive problems.

On October 15, 1962, the launchpad was used for the first time since the BLUEGILL Prime accident. Once again, the missile malfunctioned, this time approximately ninety-five seconds into launch, forcing the safety officer to destroy it. Fallout landed on the atoll.

In the following years, surveys revealed the legacy of these failed tests on Johnston Island. A 1964 check by the US Public Health Service discovered enough contaminated debris to fill four hundred barrels; these were then dumped at sea. In 1973, the Atomic Energy Commission found eight radioactive areas outside the launch site and nine on nearby Sand Island. From 1975 to 1980, more than five hundred other hot spots were discovered.[59]

## Destruction of Agent Orange

In April 1970, the deputy secretary of defense declared a suspension of the use of herbicides containing 2,4,5-T, including Agent Orange, in Vietnam. In January 1971, the final aerial spraying mission from Operation Ranch Hand flew, but smaller-scale spraying continued in Vietnam, the Korean Demilitarized Zone, Thailand, and, according to veterans, Okinawa and Guam.

The public announcement of the suspension of defoliation flights left the military with the dilemma of what to do with its remaining stocks—15,480 barrels stored at the Naval Construction Battalion Center in Gulfport, Mississippi, and the 25,000 barrels brought from Vietnam—or as the US Army report stated, from Okinawa.

Reluctant to write off these stocks, the military attempted to recoup its investment by using them as raw materials or returning them to the manufacturers; another plan involved spraying them overseas in US aid programs or domestically on government or private lands. After these proposals were rejected, the military considered dumping the chemicals into deep wells or underground nuclear test sites. As late as 1976, it considered using activated charcoal filters to remove the dioxin from the defoliants, but when it realized it had no way to safely dispose of the contaminated filters, it was forced to abandon this plan.[60]

As the military prevaricated, the barrels leaked. On Johnston Island, "approximately forty-nine thousand pounds of HO escaped into the environment annually during the period from 1972 to 1977." Authorities were so concerned about the effects of this dioxin that women were not assigned to the island. At Gulfport, contamination from the defoliant storage area spread off the base via drainage ditches, and, in 1972, the Mississippi Air and Pollution Control Commission demanded the removal of the stocks.[61]

Finally, the military resigned itself to the destruction of its arsenal of Agent Orange and decided to incinerate it aboard the Dutch ship *M/T Vulcanus*. Between July and September 1977, in a series of burns in waters near Johnston Island, the remaining herbicides were destroyed. Numerous problems occurred. Spills contaminated the crew's living quarters, and Agent Orange vapors entered the pump room. The incinerator extinguished due to water in the herbicide tanks, enveloping the ship in fumes; during one incident on July 21, an "oil-like fog was ejected from both stacks." At other times, changes in wind direction blew the plume back toward the ship. Despite these problems, US authorities continue to deny any health effects to those on board.[62] The military estimates the cost of the disposal operation exceeded $8 million. But in the years to come, extensive contamination was discovered at Gulfport and Johnson Island, and civilians and service members believe they are suffering from dioxin-related illnesses.[63]

## JACADS

In 1986, the United States began construction of the Johnston Atoll Chemical Agent Disposal System (JACADS), a prototype factory for destroying the chemical weapons brought from Okinawa in 1971, and since stored at a site on Johnston Island named the Red Hat Area. The plant was completed in 1990, with buildings to dissemble and drain munitions, incinerate the agents, and smelt their casings.

In 1990, US chemical weapons stored in West Germany were shipped to Johnston Island despite protests from environmental groups and Pacific Island governments who opposed the region becoming a military dump site. Also shipped to JACADS were mustard agent shells abandoned during World War II on the Solomon Islands, where they had since caused accidents and illnesses among the local population.

In July 1990, JACADS began operations. Already by the end of 1991, it had experienced at least six what the military refers to as "chemical events," including leaks in walls and workers mishandling VX agent. Matters grew worse in the following years. In January 1992, a VX-filled M55 rocket detonated in the kiln where it was being burned, damaging the system. In January 1993, a fire on the conveyor blazed out of control due to an inadequate sprinkler system. In March 1993, one worker was burned by mustard agent, and beginning in 1995, a series of power failures caused numerous sarin leaks into observation corridors.[64]

Rather than improving throughout time, the safety record of JACADS worsened. In 2000, there were six chemical events, at least three involving VX. The plant design allowed drip pillows placed beneath the dismantling line to become saturated with nerve agent, and following incineration, the toxin remained at levels as high as three thousand times the acceptable limits; the problem was compounded by technicians dismissing initial indications of contamination and delaying reports of the problem.[65]

By the time the decommissioning process was completed in November 2000, JACADS had destroyed two thousand tons of chemical weapons and agents. The plant had experienced thirty-nine "chemical events"; however, the number may have been higher, as the army failed to keep reports on some incidents and the commander possessed leeway on how to categorize incidents, a flaw common to self-policing.[66]

With the last of the US overseas chemical weapons destroyed, the United States now faced another problem: how to dismantle the plant and clean up its legacy of weapons of mass destruction. Between 2002 and 2004, civilian contractors worked to remove soil, consisting of 15,000 tons contaminated with dioxin and 15,000 tons contaminated with PCB and oil. At the same time as this cleanup of chemical contamination, attempts continued to deal with lingering radioactive contamination from the 1960s DOMINIC disasters. In 2002, buildings were destroyed, and cleanup teams buried "approximately 240 tons of contaminated metal debris, 200 cubic meters of concrete debris, and approximately 45,000 cubic meters of coralline soil." The landfill was then capped with sixty centimeters of soil—a method reminiscent of the Runit Dome, where at least concrete had been used. Elsewhere on the island, the JACADS main building was torn down and its debris dumped inside the bunkers that had held the Okinawa munitions; their entrances were sealed with concrete blocks.[67]

In 2003, the EPA declared the island safe, and Johnston Atoll was declared a national wildlife refuge. Tests of fish caught near the Atoll between 1995 and 2003, revealed contamination with dioxin, herbicides, radiation, arsenic, lead, and PCBs. More than a third of the fish contained 2,3,7,8-tetrachlorodibenzo para dioxin (TCDD), all contained PCBs, more than half contained traces of Americum, and 72 percent contained plutonium. In 2011, another investigation reported fish embryos near the Atoll were contaminated with PCBs.[68]

As for the human toll, today the VA denies automatic coverage for Johnston Atoll veterans suffering from Agent Orange–related illnesses; the island is not listed in the VA's guide to locations where Agent Orange was sprayed or stored, nor are veterans who supported Agent Orange incineration operations automatically eligible for help; according to the VA, "All exposures were inconsequential."[69]

Like Runit Dome, rising sea levels, storm surges, and tsunamis make Johnston Island vulnerable to swamping. Its highest point is just two meters above current sea levels, making its future destruction and dispersal of radioactive and chemical substances a problem for not only the immediate region, but also the entire Pacific Rim. Johnston Atoll is the exemplar of military ignorance and shortsightedness, a place that highlights that, while creating weapons of mass destruction is a complex and costly process, eliminating them is even more difficult.

The chemical weapons destroyed on Johnston Island were only 6 percent of the total possessed by the United States. Nine other sites on the mainland were tasked with dismantling this remaining stockpile. Initially, the military had estimated the work would cost $600 million, but the actual price tag has hit $40 billion. The process has been plagued by familiar problems: leaks, shutdowns, and cover-ups. As of 2019, the work was ongoing at two sites in the United States—Pueblo, Colorado, and Blue Grass, Kentucky. The military hopes to finally complete the work by 2023.[70]

# 9

# TOWARD ENVIRONMENTAL JUSTICE

Since 2010, when I first began reporting about military contamination in Japan, I have often been invited by academic and civic organizations to give lectures about my research. In 2016, one American university in Tokyo asked me to present in a series of public talks, so I accepted and began my preparations. A few days before the event, I was contacted by a faculty member who sounded shaken; they told me the US Embassy in Tokyo had been in touch, complaining they "didn't like my methods of doing journalism" and demanding an official from the State Department be present alongside me to rebut my talk. The university had never experienced such pressure and was upset by this attack on academic independence and freedom of expression—but they stood their ground and refused to change the event. Next, the State Department took its complaint to the university's headquarters in the United States; the institute there relented and instructed its Tokyo branch to schedule a lecture for the Embassy so it could refute what I had to say.

At the university, I gave my talk to a standing-room-only audience. I explained the history of Japanese discrimination against Okinawa, the island's suffering during World War II, and the environmental degradation caused by the military in the years since. The attendees asked incisive questions, and they included young US veterans who were concerned about how their own service on Okinawa may have impacted their health. After my talk, the university distributed flyers advertising the following month's State

Department counterlecture, billed as the "role of the United States in the security of Okinawa and the rest of Japan."

Several days later, the university contacted me with some surprising news: the American Embassy had canceled its talk, claiming staff were too busy to provide a speaker. Immediately I filed a Freedom of Information Act (FOIA) request with the State Department to find the background on the decision. Almost one year later, I received the records, and the internal e-mail correspondence revealed the real reason for its cancelation: "This sounds like a death trap," wrote the Embassy's political-military director.

There are three lessons to be learned from this incident. First, it shows the lengths to which the US government tries to cover up its contamination of Japan, to the extent that it is even willing to interfere with free speech and obstruct open discussion on the topic. Second, the State Department's cancelation of its own talk displays its aversion to public accountability in Japan, a nation Washington repeatedly claims is its number-one ally. Most importantly, the incident proves the State Department is fully aware of how severely its military is damaging Japan's environment; the "death trap" comment was made by the embassy's political-military director, whose office sends a representative to US–Japan Joint Committee meetings. When it comes to recognition of the US poisoning of Japan, opacity reigns, and it extends to the very top of the US–Japan alliance.

## SEVENTY-FIVE YEARS OF POISONING

Okinawa, Guam, Saipan, Tinian, and the Marshall Islands share a common history. Each one experienced Japanese brutality, including cultural repression, forced labor, rape, and mass killing. Each one was liberated by US forces, which lost thousands of their own men during the most intense combat in the Pacific. After the war, islanders looked forward to enjoying their newfound freedoms, but instead they found themselves trapped in geopolitical limbo without the democratic or constitutional rights afforded full-fledged states or independent nations. These gray-zone statuses enabled the military to seize and occupy their best agricultural and fishing areas, and displace large percentages of their populations; politically powerless, residents were unable to prevent the United States from using their islands as sandboxes to test weapons of mass destruction. As well as radiation, dioxin, chemical, and biological weapons, their communities were polluted with hundreds of other toxins—fuel, pesticides, polychlorinated biphenyls (PCBs), trichloroethylene (TCE), per- and polyfluoroalkyl substances (PFAS), asbestos, and lead—which, while less headline-grabbing than nerve agent and thermonuclear bombs, have taken their toll on the health of local peoples.

All too often, children have borne the brunt of military contamination: the jellyfish babies and thyroid-sickened youngsters of the Marshall Islands, and the Okinawans exposed to CS gas in their classrooms, burnt while swimming and sickened by herbicide leaks into water supplies. Also impacted have been the children of service members exposed to water pumped through lead pipes and dioxin-contaminated soil near their schools. The young are particularly susceptible to environmental pollutants. Small volumes of toxins that might have minor effects on fully grown adults can have a serious impact on children, whose body weight is much lower. Such substances as dioxin, lead, and PFAS can build up in young bodies and disrupt their development physically and mentally.

Impacted, too, have been hundreds of thousands of adults. These include Marshallese and Guam downwinders exposed to nuclear fallout; the Japanese fishermen doused in radiation following the Bravo test and sickened by Okunoshima's chemical weapons dumped at the end of World War II; civilians consuming water so contaminated with fuel it could be set alight; those made ill by aircraft noise; and those whose fields and fishing waters have been polluted with sewage, lead, and herbicides.

On the bases, American service members and local workers have been poisoned throughout the Pacific region. Unprotected by the health and safety regulations of civilian industries, they have been exposed to radiation while attempting to clean up the guinea pig fleet, sickened by leaks of chemical weapons, and soaked in herbicides on their way to and from Vietnam. They and their families have been exposed to drinking water contaminated with TCE and other poisons. The Japanese government has treated its own World War II workers with equal disregard, refusing for years to help those poisoned while manufacturing chemical weapons on Okunoshima and, more recently, denying help for those exposed to asbestos while on US bases.

For as long as the US military has been sickening people in the Pacific, it has also been doing all it can to downplay and conceal its contamination. Starting with the bombings of Hiroshima and Nagasaki, where it said radiation exposures had been minimal, it then extended its cover-up to postwar nuclear tests in the Marshall Islands and overseas. During the Cold War, military spokesmen, diplomats, and the CIA dubbed its critics "anti-American," lied about the extent of contamination, and stonewalled investigations.

Both the United States and Japan have abandoned the victims of their lost wars. The United States has not compensated any of the 3 million Vietnamese and uncountable Laotians and Cambodians exposed to dioxin-tainted herbicides; likewise, Japan has ignored calls for justice for the hundreds of thousands of Chinese harmed by its biological and chemical weapons during and after World War II. The United States must also shoulder some blame: By offering immunity to Unit 731 scientists and then locking away their research, it let Tokyo deny knowledge of their operations, avoid helping Chinese survivors, and keep the Japanese public from reflecting on its biochemical war crimes on mainland Asia, a policy of national amnesia abetted by the United States during its occupation.

The deal the United States struck with Japanese military scientists and the release of other suspected war criminals forged a rotten core at the foundation of the US–Japan military alliance: Both countries were complicit in the use of weapons of mass destruction to slaughter civilians. Yet, neither nation has ever truly reflected on its actions. This set the precedent for how Japan dealt with the United States in the coming years: It gave priority to the US military over the rights of its own citizens. Unit 731 scientists moved seamlessly from abusing Chinese civilians to abusing Japanese *hibakusha*—with the goal of developing more lethal weapons of mass destruction. Japan turned its back on its own radiation-sickened fishermen to acquire nuclear power, which, with the help of the CIA, it convinced the public was a safe source of electricity. The Japan–US Status of Forces Agreement (SOFA) enshrined the right of the US military to pollute Japan with impunity; Japanese taxpayers have spent billions of yen cleaning up contaminated land, and Tokyo has never attempted to renegotiate the arrangement.

US service members and their families have suffered from this complacency. Without the protection of Japanese laws, they have been at the mercy of their own self-policing military, which has let them drink contaminated water and live on bases that used to store Japanese chemical weapons but have never been thoroughly surveyed for contamination. At the same time, the Japanese government is failing to protect the American men and women stationed there under the Security Treaty. At Naval Air Facility Atsugi (NAFA), Japanese leaders at a local level cast doubts on the contamination caused by the Shinkampo incinerator, while the national government stood by and did nothing. In the wake of Operation Tomodachi, the Japanese government has turned its back on US service members who sacrificed their health to help Tohoku residents to recover from the nation's worst postwar disaster.

In Japan, the government's failure to hold the United States accountable for its environmental transgressions has most harmed Okinawa. When Okinawa was returned to Japanese control in 1972, Tokyo promised to reduce the military presence there to the same as that on the mainland; it has broken this pledge, and the prefecture still hosts 70 percent of military facilities. Instead of protecting its newly returned prefecture from environmental contamination, Japan instead has allowed the United States to continue polluting it; it has never updated SOFA, it hid the 1973 agreement allowing on-base checks for decades, and Prime Minister Abe Shinzo's administration has been even more subservient than preceding Liberal Democratic Party (LDP) governments. The inevitable outcome of American and Japanese negligence and collusion has been the contamination of Okinawa's drinking water with cancerous PFAS, a problem likely to impact islanders for a generation. Tokyo's withholding of on-base PFAS test data epitomizes this disdain.

Such treatment of Okinawa falls into a long pattern of discrimination. In the spring of 1945, 140,000 islanders were sacrificed to protect the mainland and the imperial system. In 1952, with the blessing of the emperor, Okinawa was severed from the nation and suffered greatly during the Vietnam War. In 1972, the island again became part of Japan, but the bases remained. Today, Tokyo still refuses to recognize Okinawans as an indigenous people, preferring to assert the nation is one homogenous group. Such structural discrimination feeds into its ongoing suppression of democracy on the island; consecutive elections and mass demonstrations have demanded a reduction in the number of bases. It is impossible to justify such a disproportionate military presence: Financially, the bases occupy 15 percent of Okinawa's main island but contribute only 5 percent to the prefectural economy. Militarily, packing so many facilities into such a small area makes them vulnerable to attack or natural disaster. Morally, the prereversion promise must be honored and the base burden shared throughout Japan so all residents can understand the dangers of military operations conducted in their backyards.

The United States has treated its territories with similar contempt for human rights, giving its military the freedom to trample indigenous voices, seize land, and ignore calls for the preservation of historic sites. Prior to World War II, the United States attempted to stamp out Chamorro, even burning books written in the language. During the war, residents hid US troops, at great risk to their own safety. As the US military approached, they rose up against the Japanese occupiers. Today, Chamorros enlist at a higher rate than mainland Americans, but despite this, the federal government denies their constitutional rights; such lack of representation in Washington enables the United States to ignore demands for World War II reparations and coverage according to the Radiation Exposure Compensation Act (RECA).

Nowadays, the Pentagon likes to claim it is a responsible steward of the environment; however, in the Pacific, it has only moved to remediate contamination when it has been forced to do so, for example, on Runit Island, where residents threatened to sue the United States—still then the subsequent cleanup work was perfunctory. Further improvements in the years to come will likely stem from self-interest as well. On Okinawa, the eventual closure of Marine Corps Air Station (MCAS) Futenma will win the United States huge public relations points and, at the same time, protect it from the potential liability of the cleanup, which will cost billions of dollars. On Guam and the Commonwealth of the Northern Mariana Islands (CNMI), the United States may agree to better environmental controls to sell a skeptical public on its expanded presence in the region. Likewise, any future help for the Marshall Islands will likely only materialize to guarantee the ongoing use of the world's largest lagoon at Kwajalein as a target for rockets fired from California.

The United States has failed to learn from its decades of mistakes; its grotesquely bloated budget means it doesn't need to. Japan also refuses to heed the lessons of the past. Despite the nuclear attacks on Hiroshima and Nagasaki, the 1954 exposure of its fishing vessels, and the 2011 Fukushima meltdowns, it continues to promulgate the use of nuclear power. The LDP government and Nuclear Village force families to return to their contaminated properties and allow power plants to operate near active fault lines; Prime Minister Abe Shinzo lied to the world when he declared the meltdowns at Fukushima Daiichi "under control," even as thousands of tons of radioactive water poured into the Pacific.

Today, the Japanese government refuses to sign the 2017 United Nations Treaty on the Prohibition of Nuclear Weapons, and when the International Campaign to Abolish Nuclear Weapons (ICAN) won the Nobel Peace Prize in 2017, Abe ignored requests to meet its representatives. As it always has done, the Japanese government slavishly backs US policies on nuclear weapons and supports the supposed protection proffered by its nuclear umbrella. At the same time, conservative politicians tout Japan's ability to create its own nuclear devices. Its power stations have provided it with enough plutonium to make six thousand warheads; Japan is just a "screwdriver's turn" from producing its own nuclear weapons.

In the years to come, the militaries of Japan and the United States will work together ever more closely thanks to Abe's steamrolling the State Secrets and Collective Self-Defense legislation through parliament. It is a frightening proposition; neither nation has ever reflected on its World War II use of weapons of mass destruction, nor have they learned any lessons of their long-term impact in terms of contamination, cleanup costs, or damage to human health. Both Japan and the United States have allowed the needs of the military to take precedence over civilian health, and the more the two nations cooperate, the worse the problem will become.

## ENVIRONMENTAL JUSTICE

Against this bleak backdrop, there are signs of hope. Across the Pacific, as people become aware of the military's environmental damage to their communities, a wave of resistance has developed. These movements have crossed traditional barriers, bringing together antibase activists, military veterans, whistleblowers, and indigenous rights groups united by the common goals of safeguarding human health and the environment, securing land rights, and holding the Pentagon accountable for its injustices.

The Japanese movement traces its origins to the mass-petition drive following the *Lucky Dragon* disaster, whereby a third of the population called for an end to nuclear tests; more recently, Tokyo-based Peace Boat and other non-governmental organizations have supported ICAN and large crowds have turned out to protest Japanese government plans to restart nuclear reactors shuttered in the wake of the Fukushima Daiichi meltdowns. On mainland Japan, such citizens' groups as Rimpeace have been using FOIA to monitor US military environmental infractions, and their research has uncovered serious problems at many of the worst-offending bases. Japanese lawyers have been active in representing the Chinese victims of their nation's chemical and biological weapons to convince their own government to admit to their use and compensate survivors.

On Okinawa, Professor Sakurai Kunitoshi, former president of Okinawa University, has long urged the US military and the Japanese government to be more accountable for environmental damage. Moreover, for many years, Kawamura Masami served as director for Citizens' Network for Biodiversity in Okinawa, guiding projects to win transparency on plans for the new United States Marine Corps (USMC) base at Henoko and independent cross-checks on work at the dioxin-contaminated soccer pitch in Okinawa City. In 2016, Kawamura established the Informed-Public Project, which

combines on-the-ground research with extensive use of information dis-
closure requests to investigate how Japanese government complacency
compounds US military environmental problems. Unlike elsewhere in Japan, Okinawa prefectural officials have been proactive in handling military
contamination; in 2017, Okinawa became the only prefecture to compile
environmental information for US bases—including past accidents and potentially impacted water sources—and upload the data for public access in
the closest thing Japan possesses to the online system maintained stateside
by the Environmental Protection Agency (EPA).

The Marshallese have a long history of demanding environmental justice
from the United States. John Anjain, a Rongelap magistrate whose son died
from leukemia related to radiation exposure, documented the illnesses of
his community and led calls for the United States to provide compensation.
Jeton Anjain, a Marshallese senator, helped residents move from Rongelap
in 1985, following revelations the island was still severely contaminated,
and he pressured the US government to conduct radiation checks. Tony
deBrum witnessed the Bravo explosion as a nine-year-old, and as an adult,
he helped negotiate independence for the Marshall Islands and served as a
government minister; as well as speaking out against nuclear weapons, he
was also an outspoken activist on climate change, which has already begun
to swamp the low-lying Marshall Islands.

On Guam, Angel Santos led calls for environmental justice for the Chamorro people. The United States Air Force (USAF) veteran had lost a young
daughter to cancer linked to the TCE contamination in the drinking water
at Andersen Air Force Base, where he had been stationed. He became
one of the founding members of Nasion Chamoru, participating in sit-ins
on ancestral land seized by the military; as a Guam Legislature senator, he
continued to fight for environmental justice and the return of property to
the Chamorro people.

A new generation of Chamorro activism emerged in response to Pentagon plans to escalate military use. We Are Guahan organized the reading
of the ten thousand pages of the 2010 draft environmental assessment and
highlighted its key problems, particularly the planned firing range at Pagat.
Together with the National Trust for Historic Preservation, they sued the
Pentagon, and as a result, the plan was abandoned. Today, Prutehi Litekyan
(Save Ritidian) works to block plans to build a firing range in the pristine
forest there and cooperates with activists throughout the region. Army
veteran Robert Celestial, a Runit Dome participant and president of the
Pacific Association of Radiation Survivors, has been fighting tirelessly to en-

sure that the military is held accountable for the fallout that contaminated Guam following the 1952 IVY Mike blast.[1]

Elected representatives have also been aggressively holding the military accountable. This includes Therese Terlaje, who has supported justice for those exposed to dioxin herbicides and radiation; Sabina Perez, a central figure in Prutehi Litekyan; and Attorney General Leevin Camacho, who has been active in dealing with PFAS contamination on the island and maintaining pressure on the Navy to fund some of the Ordot dump cleanup.

In CNMI, plans by the Pentagon have triggered a large backlash. Under the umbrella organization Alternative Zero Coalition, grassroots groups have united to dissect the flaws in the Department of Defense's draft assessments for the buildup, and with the help of the nonprofit Earthjustice, they have filed legal action against the military. In March 2019, CNMI began offering homestead lots on Pagan for people to return to the island, a move that would hamper Department of Defense plans to bomb there.[2]

Some critics have accused these groups of being "antimilitary," but they fail to acknowledge that many of those fighting the hardest for environmental justice are themselves former service members. Veterans of Okinawa and Guam have harnessed social media to gather accounts of herbicide usage on the islands, sharing advice and testimony to piece together the jigsaw of the chemicals' usage. A similar approach raised public awareness of the contamination at NAFA, Runit Island, and Johnston Atoll.

Whereas the national media often ignores such grassroots movements, the local media works hard to uncover and report on the military's environmental damage. On Okinawa, both daily newspapers, *Ryukyu Shimpo* and the *Okinawa Times*, where I report as a special correspondent, and the TV station Ryukyu Asahi Housou understand the importance of an assertive, investigative media to wring transparency from Japanese and US authorities. In Micronesia, too, the media realizes that publicizing the Department of Defense's environmental damage is not antimilitary or unpatriotic, and it regularly reports on contamination and publishes editorials calling for the military to be held accountable for its actions. In CNMI, the *Saipan Tribune* was instrumental in chronicling the PCB debacle in Tanapag, and today it is not afraid to condemn the Pentagon's planned buildup. With local newsrooms throughout the world laying off staff and newspapers disappearing, this coverage proves just how vital local news is to the functioning of both healthy democratic debate and human health itself.

In the Pacific, tens of thousands of people have joined demonstrations and engaged in direct nonviolent action, risking liberty and safety to protest the military's destruction of their islands' environments. The 1969 leak of nerve agent kindled mass demonstrations on Okinawa, forcing the eventual removal of the arsenal; more recently, plans to landfill Oura Bay with a new USMC base have led to one of the longest sit-ins in the world, with residents taking to the sea in canoes to block construction since 2004. During these demonstrations, attended by tens of thousands of Okinawans, speakers decry the storage of weapons of mass destruction on their island, their accidents, and their potential to indiscriminately kill millions.

From 1969 to the 1980s, Marshallese took to the seas to stage sail-ins and sit-ins, whereby they moved onto their home islands, which were being used by the military. They blocked at least one missile test and succeeded in winning some concessions—albeit limited—from the United States, including infrastructural improvements and increased lease payments.

The spread of the internet has allowed Pacific communities to forge connections with other indigenous groups fighting to preserve their lands, including the movement to block the Dakota Access Pipeline on Sioux tribal property and the construction of the Thirty Meter Telescope on sacred land at Mauna Kea, Hawaii. Okinawans have also taken inspiration from Puerto Ricans who managed to convince the military to withdraw from Vieques island, which it contaminated in ways similar to Okinawa. Such connections help people to better understand their roles within the context of the wider fight against injustices; they can share tactics to defend their water and land, while gaining a sense of solidarity in the daunting struggle to confront the largest military the world has ever seen.

## SOME GUIDING PRINCIPLES

Current US policy toward past and present contamination is mired in opacity, irresponsibility, and arrogance. Deeply entrenched in the military is a culture of secrecy and entitlement that has allowed it to act with zero oversight for decades. On the US mainland, the federal government and local authorities have pushed the military to become more transparent, resulting in some improvements; however, in the Pacific region, due to the geographic distances, powerlessness of local governments, and extreme toxicity of the substances involved, local communities continue to suffer from the environmental impact of military operations. In coming years, with Department of Defense plans to expand in the region, the problem will continue to grow unless there are fundamental changes. What is needed is a new approach anchored in the principles of accountability, transparency, and human rights.

## Accountability

First and foremost, Japan and the United States must be held accountable for their use of nuclear, chemical, and biological weapons during and after World War II. When China and Japan signed their peace treaty in 1978, records cataloging Unit 731 biological weapons use were locked in the United States, and people were unaware of the more than four hundred thousand chemical munitions abandoned in China. Now that the extent of these operations is better known, compensation must be provided for any remaining survivors and their families. Similarly, when the Marshall Islands became independent in 1979, and the Nuclear Claims Tribunal was established, the long-term persistence of radioactive contamination was not well understood; now we know the region remains dangerous and some places may forever be uninhabitable. The United States needs to recognize this damage and honor outstanding claims, taking full responsibility for the destruction it caused. Guam, too, ought to be treated as equally as stateside downwind areas; as recommended by the National Academy of the Sciences, the island needs to be covered by RECA forthwith. Soldiers who took part in decontamination work at the Runit Dome and Johnston Atoll ought to be eligible for automatic VA coverage; today the United States helps troops exposed directly after nuclear tests—the so-called Atomic Soldiers—and to deny help to the men who cleaned up in later decades is unjustifiable. Such compensation will likely cost billions of dollars, but if this forces the Pentagon to grasp the economic consequences of manufacturing such weapons of mass destruction, it may be less inclined to keep making them.

Since the 1990s in the United States, federal regulations have compelled the military to clean up its bases, but in Japan, no such work has started. The problem is particularly urgent on Okinawa, where the concentration of bases has also concentrated contamination. Without a fundamental overhaul of SOFA, it will be impossible to halt the recklessness with which the US military operates in Japan. Self-regulation has proven a failure—the military cannot be depended on to police itself, so accountability must start at the individual level. Service members whose actions cause off-base damage ought to be held accountable by Japanese courts in the same way Japanese authorities have jurisdiction over US troops committing crimes against Japanese citizens today. If breaches are serious enough, such accountability must extend as high as the base commander. In the United States, prosecutions of service members for environmental infractions made superior officers understand they were not above the law; those in Japan need to feel the same weight of personal responsibility.

The principle of accountability must extend to past contamination. The military ought to conduct thorough tests and make the results public in a timely manner. These checks must be supervised by independent specialists; experience has taught us that neither the US military nor Japanese authorities can be trusted. When contamination is found, the military should be held responsible for its remediation; the polluter ought to pay. Concerning contamination on returned land, the US government must be held financially responsible for cleanup costs, damage due to delayed reuse, and health checks for those exposed to dangerous substances.

## Transparency

The US government's decision to grant immunity to Japanese scientists who committed atrocities in China was abhorrent, making the United States complicit in these war crimes. It is not too late to make amends. The Japanese and US governments must immediately release full records of the Japanese scientists involved in human experimentation and the work they performed; they need to be named, shamed, and posthumously stripped of their academic credentials. As recently as 2018, Kyoto University still allowed one Unit 731 scientist to retain his degree based on a thesis he had written drawing on human experiments; the scientist had changed the words *humans* to *monkeys* but had not eliminated references to the test subjects complaining of headaches. Permitting such suspected war criminals to retain their qualifications is a stain on the reputation of these universities.[3]

The United States, too, must be forced to open its records on historic weapons of mass destruction trials on Okinawa. Army reports state the island was a Project 112 site. Yet, it is not included in the Department of Defense's official list of locations, denying veterans justice. Moreover, there needs to be a thorough investigation into service members' accounts that such herbicides as Agent Orange were deployed to Okinawa. According to the 2018 Government Accountability Office (GAO) report about herbicides on Guam, the Department of Defense decided to initiate surveys there despite the "low probability of conclusively identifying the components of Agent Orange" to "address veterans' and the public's concerns." Okinawa must be granted the same respect; hundreds of veterans are seriously ill, Okinawa's off-base communities were exposed, and the land has been contaminated. Dr. Alvin Young's 2013 investigation failed to interview veterans or conduct tests, raising serious questions about its reliability. Veterans, their families, and Okinawans deserve a full and fair enquiry.[4]

Today in Japan, the seventy-eight US bases are black holes in which no one—not even installation commanders—knows what dangers exist. This must change. The US military must release maps of dump sites, hazardous material storage areas, and their inventories for both closed and (security concerns permitting) active bases. This information must be available to local communities so in the case of on-base fires, they are able to draw up effective plans to extinguish blazes and evacuate residents in potential danger.

Documents alone are not enough. Record-keeping during the Cold War was remiss. Thus, interviews with former base workers and US veterans must be conducted to fill in the gaps in our understanding. Such a combination of documentary evidence and oral testimonies helped Japan's 2003 survey of chemical weapon dump sites, and it might be used as a model for surveys of postwar US contamination.

The PFAS problem shows that the damage to Okinawa is urgent—and Tokyo's failure to share on-base checks makes it a coconspirator. The terms of the 1973 Joint Committee memorandum need to be honored immediately. To ascertain whether bases are damaging civilian communities, local, prefectural, and national authorities need to be allowed to access installations, inspect sources of contamination, and, if necessary, take soil, air, and water samples. Likewise, US service members should be afforded the right to request similar on-base inspections without fear of repercussions.

With the principle of transparency in mind, in 2019, I donated more than five thousand pages of documents to Okinawa International University's Jon Mitchell Collection; obtained via FOIA, the reports detail contamination on US bases in the prefecture from 1945 to 2017. To facilitate veterans' claims with the VA and help their doctors understand their potential exposure, I have uploaded some of the key documents to the publisher's homepage. Also, on my own website—www.jonmitchellinjapan.com—there are testimonies from service members who believe they were sickened while serving in Japan.

## Human Rights

At stake today is the violation of human rights on an unprecedented scale, stretching across hundreds of thousands of square kilometers and more than seven decades. The right to a clean environment is fundamental—the right to know whether the land you live on is safe or whether it is poisoning the health of yourself and your children, the right to know if the air you're breathing is free from radiation, nerve agent, or dioxin, the right to know if the water you're swimming in, fishing in, showering in, and drinking is exposing you to substances that cause cancer. The US military is violating these very basic human rights.

Washington likes to take credit for the maintenance of peace in the postwar Pacific, but it has contaminated communities here worse than many war zones and plans to continue doing so at an increased rate in the coming years. There needs to be an overhaul of this mindset.

Today, the US military has an annual budget of approximately $700 billion a year, roughly 15 percent of the federal total and taking more than 50 percent of discretionary spending. Its destruction of chemical weapons has cost $40 billion. Dismantling nuclear aircraft carriers will require $1 billion per vessel. The crash of a single B-2 on Guam in 2008 was another $1.2 billion. In Japan, taxpayers spend 190 billion yen ($1.7 billion) per year hosting US military bases; we are paying billions more to build a new USMC base at Henoko, ignoring both the popular will of the Okinawan people and the warnings of geologists who say the seabed at the site is as soft as mayonnaise.

These costs are obscene. On a national level, the United States and Japan are suffering from urgent problems: The United States has crumbling infrastructure and huge income disparity, while Japan has a poverty rate of 16 percent and a broken pension system that, in the coming years, will plunge millions of retirees into hardship. For both nations, the economic problems are even worse on their Pacific islands. Okinawa is one of the poorest prefectures in Japan, with salaries, household savings, and child poverty consistently among the worst in the nation. With its best land occupied by military installations, housing is cramped and roads are congested, and it remains the only prefecture without an intercity train network. Guam has a poverty rate of 23 percent, higher than any US state, whereas in the CNMI, the figure is a staggering 52 percent. The now-independent Marshall Islands has a per person gross domestic product (GDP) of $3,600, and overcrowding is worsening because islanders are unable to return to some regions due to radioactive contamination.

The health crises of these islands are well-documented. In the Marshall Islands, serious illnesses are "extreme"; Guam is beset by diabetes and cancer; and on Okinawa, once so famed for its long-living residents that a diet was named after it, the mortality rate for those younger than sixty-five has recently become the worst in the nation. While many studies blame these statistics on lifestyle-related diseases, they ignore the impact of military occupation; residents did not willingly choose to consume processed foods.

They have been forced to do so by the seizure and subsequent contamination of their prime agricultural and fishing areas.[5]

Also neglected by researchers has been the extent to which military contamination has exacerbated islanders' health problems. To date, there have only been piecemeal surveys, the results of which have been frightening: high levels of PFAS in residents living near MCAS Futenma; infant mortalities on Guam, where veterans claimed to have sprayed dioxin herbicides; and high cancer rates near PCB-polluted Cocos Lagoon. The US and Japanese governments have not funded follow-up surveys, likely based on fear that any discovery of widespread health problems linked to military contamination would trigger an unprecedented backlash, undermining the Pentagon's presence throughout the region.

So how many people have been exposed? Table 9.1 attempts to chart the scope of postwar US military contamination in the western Pacific. It omits the victims of radiation poisoning who died in the years following the attacks on Hiroshima and Nagasaki; likewise, it does not include the millions on mainland Asia exposed to Japanese chemical weapon and US dioxin. Two other caveats ought to be kept in mind: 1) For some of the incidents, for example, the contamination of Kadena's fiery well water and those who consumed tainted vegetables, no estimates exist, and 2) the military has undoubtedly kept some incidents under wraps.

**Table 9.1. Estimated Human Impact of Military Contamination in Japan, Okinawa, and the Micronesia Region**

| Year | Event | Location | Numbers Exposed |
|---|---|---|---|
| 1946 | Operation CROSSROADS | Marshall Islands | 42,000 sailors |
| 1952 | fuel and oil in water supply | Tachikawa Air Base, Japan | 10,000 civilians |
| 1952 | Operation IVY Mike | Marshall Islands/Guam | 67,000 on Guam |
| 1954 | Operation CASTLE Bravo | Pacific Ocean | 13,000 Marshallese, 1,000 fishermen |
| 1960s–1980s | herbicide use | Okinawa | >250 service members |
| 1968 | herbicide use | Gushikawa, Okinawa | 230 civilians |
| 1968 | fuel leak | Futenma, Okinawa | 1,000 civilians |
| 1969 and 1971 | nerve agent leak | Okinawa | 26 service members, one US civilian |
| 1970s onwards | PFAS contamination of water supply | Yokota Air Base, Japan | 11,500 service members and dependents |
| 1970s onwards | PFAS contamination of water supply | Okinawa | >450,000 civilians; tens of thousands of service members, and tourists |

(continued)

**Table 9.1.** **(continued)**

| Year | Event | Location | Numbers Exposed |
|------|-------|----------|-----------------|
| 1973 | chemical spills | Okinawa | 180 civilians |
| 1977–1980 | dumping of radioactive waste | Runit Island | 4,000 soldiers |
| Late 1970s | TCE contamination of water supply | Andersen Air Force Base, Guam | 1,200 service members and dependents |
|  |  | Total | >601,000 |

Many of these 600,000 people have been exposed to numerous toxins, which can accumulate throughout a lifetime and interact with one another to trigger serious illnesses. Civilians living near bases have been exposed for longer periods than service members and dependents whose tours last only a few years. Children are more susceptible to contamination than adults.

Human health must take precedence over everything else, and the failure of governments to investigate the impact is criminally negligent. Local and national authorities in the Pacific must create a health registry to track illnesses in their populations, identify spikes near past and present military installations, and assist those affected. No cost must be spared; there is always more money for new weapons and wars but never enough for those poisoned by past ones. Too many people have been exposed for too long. *Nuchi du takara*, goes a traditional Okinawa saying: *Life is a treasure and more important than anything.*

# APPENDIX

## Contaminants

## 2,4,6-TRINITROTOLUENE (TNT)

2,4,6-trinitrotoluene (TNT) is a yellow, odorless solid used as an explosive in military shells, bombs, and grenades. Inhalation of high concentrations of TNT in the air can cause such harmful health effects as anemia and abnormal liver function. Other effects include skin irritation after prolonged skin contact and cataract development after long-term exposure. The Environmental Protection Agency (EPA) has determined that 2,4,6-trinitrotoluene is a possible human carcinogen.

## ARSENIC

Arsenic is a gray, metal-like material that occurs naturally; inorganic arsenic compounds have been used as wood preservatives and pesticides. Inhalation of these compounds may cause respiratory irritation, nausea, skin effects, and increased risk of lung cancer. Oral exposure causes such symptoms as nausea, vomiting, diarrhea, cardiovascular effects, and encephalopathy (disturbances of brain function). There may also be an increased risk of skin, bladder, and lung cancers.

## ASBESTOS

Asbestos is the name given to a group of six different fibrous minerals that have been used in building materials, friction products (e.g., automobile brake parts), and heat-resistant fabrics. Breathing high levels of asbestos fibers for a long period of time may result in scar-like tissue in the lungs and the pleural membrane that surrounds the lung, leading to a disease called asbestosis. Asbestosis is a serious disease and can eventually lead to disability and death. Asbestos is a carcinogen.

## BENZENE

Benzene, a colorless liquid with a sweet odor, is found in gasoline and other fuels. Breathing high levels of benzene can result in death, while lower levels can cause drowsiness, dizziness, rapid heart rate, headaches, tremors, confusion, and unconsciousness. The major effect of benzene from long-term exposure is on the blood. Benzene causes harmful effects on the bone marrow and can cause a decrease in red blood cells, leading to anemia; it can also cause excessive bleeding and affect the immune system, increasing the chance for infection.

## BORON

Boron is an element from which compounds—called borates—are used in the production of glass, fire retardants, pesticides, and high-energy fuel. Exposure to large amounts of boron during short periods of time can affect the stomach, intestines, liver, kidneys, and brain, and can lead to death. Animal studies indicate that the male reproductive organs, especially the testes, are also affected.

## CADMIUM

Cadmium is a natural element usually found as a mineral combined with other elements. It has many uses, including in batteries, pigments, metal coatings, and plastics. Breathing high levels of cadmium can severely damage the lungs; consuming high levels severely irritates the stomach, causing vomiting and diarrhea. Long-term exposure to lower levels of cadmium leads to a buildup in the kidneys and possible kidney disease; other long-term effects are lung damage and fragile bones. Cadmium and its compounds are carcinogenic.

## CHLORDANE

Chlordane was used as a pesticide in the United States from 1948 to 1988. Chlordane affects the nervous and digestive systems, and the liver. Headaches, irritability, confusion, weakness, vision problems, vomiting, stomach cramps, diarrhea, and jaundice have occurred in people who breathed air containing high concentrations of chlordane or swallowed small amounts. Large volumes of chlordane taken by mouth can cause convulsions and death.

## CHROMIUM

Chromium is a naturally occurring element that has several compounds, one of which is chromium (VI)—hexavalent chromium—which is used for chrome plating and wood preserving. Breathing high levels of chromium (VI) can cause irritation to the lining of the nose; nose ulcers; runny nose; and such breathing problems as asthma, cough, shortness of breath, or wheezing. Inhalation can lead to lung cancer, and ingestion can cause stomach tumors.

## DDT

DDT (dichlorodiphenyltrichloroethane), a white, crystalline solid with no odor or taste, is a pesticide once widely used to control insects in agriculture and insects that carry such diseases as malaria. Its use in the United States was banned in 1972, because of damage to wildlife. DDT affects the nervous system. People who accidentally swallowed large amounts of DDT became excitable and had tremors and seizures. These effects went away after the exposure stopped. Commercial DDT can also be contaminated by the similar substance DDE (dichlorodiphenyldichloroethylene). One study showed that women with high amounts of DDE in breast milk had an increased chance of having premature babies.

## JET FUELS (JP-4 AND JP-7)

Jet fuels JP-4 (jet propellant-4) and JP-7 (jet propellant-7) are flammable, colorless to straw-colored liquid mixtures made according to US Air Force standards for use as aircraft fuels. Inhaling large amounts of JP-4 vapor may cause painful breathing and a feeling of suffocation, as well as headache, dizziness, nausea, depression, anxiety, memory loss, and irritability.

## JET FUELS (JP-5 AND JP-8)

JP-5 (jet propellant-5) and JP-8 (jet propellant-8) are colorless, kerosene-based fuels used in military aircraft. Little is known about the effects of JP-5 and JP-8 on people's health. Results from a few studies of military personnel suggest that exposure to JP-8 can affect the nervous system; some effects observed include changes in reaction time and neurological function.

## LEAD

Lead is a bluish-gray metal used in the production of batteries, ammunition, metal products (solder and pipes), and devices to shield X-rays. The effects of lead are the same whether it enters the body through inhalation or ingestion. Long-term exposure can result in decreased learning, memory, and attention, and weakness in fingers, wrists, or ankles. Lead exposure can cause anemia and damage to kidneys. It can also cause an increase in blood pressure, particularly in middle-aged and older individuals. Exposure to high lead levels can severely damage the brain and kidneys, and cause death. In pregnant women, exposure to high levels of lead may cause miscarriage; high-level exposure in men can damage reproductive organs.

## LEWISITE

Lewisite is an oily, colorless liquid with an odor like geraniums. Inhalation irritates airways, causing burning pain in the nose and sinuses, laryngitis, cough, shortness of breath, nausea, and vomiting. It can also damage airway tissues and lead to accumulation of fluid in the lungs, resulting in death. Contact of the skin with lewisite will result in local pain, swelling, and rash, followed by blistering that might be delayed for hours. Contact with the eyes causes immediate pain, rapid swelling, and serious damage to the cornea and other parts.

## MALATHION

Malathion is a brownish-yellow insecticide that smells like garlic. It interferes with the normal function of the nerves and brain. Exposure to high levels of malathion for a short period of time in air, water, or food may cause difficulty breathing, vomiting, diarrhea, blurred vision, sweating, headaches, dizziness, loss of consciousness, and death.

## MERCURY

Mercury is a shiny, silver-white, odorless metal that combines with other elements to form a variety of compounds. Short-term exposure to high levels of metallic mercury vapors may cause lung damage, nausea, vomiting, diarrhea, an increase in blood pressure or heart rate, skin rashes, and eye irritation. Exposure to high levels of metallic or mercury compounds can permanently damage the brain, kidneys, and developing fetus. Effects on brain functioning may result in irritability, shyness, tremors, changes in vision or hearing, and memory problems.

## MUSTARD AGENT (HD)

Mustard agent is typically a yellow to brown oily liquid with a slight garlic or mustard odor. It burns skin and causes blisters within a few days of exposure; contact with the eyes causes them to burn, eyelids to swell, and repeated blinking. Inhalation of mustard agent causes coughing, bronchitis, and long-term respiratory disease; exposure to large amounts can be fatal. Long-term exposure may lead to cancer of the upper respiratory airways.

## NERVE AGENTS (SARIN AND VX)

Sarin (GB) is the most volatile nerve agent, whereas VX, an oily liquid, is the least volatile. Manifestation of nerve agent exposure includes runny nose, chest tightness, pinpoint pupils, excessive salivation and sweating, vomiting, involuntary defecation and urination, seizures, paralysis, coma, and death. Fatigue, irritability, nervousness, and memory defects may persist for as long as six weeks after recovery from exposure.

## PENTACHLOROPHENOL

Pentachlorophenol was widely used as a pesticide and wood preservative, but since 1984, its purchase and use has been restricted to certified applicators. Exposure to high levels of pentachlorophenol can cause the cells in the body to produce excess heat, leading to a high fever, profuse sweating, and difficulty breathing. Body temperature can increase to dangerous levels, causing injury to various organs and tissues, and even death. Long-term exposure can also cause liver effects and damage to the immune system. It is a probable human carcinogen.

## PER- AND POLYFLUOROALKYL SUBSTANCES (PFAS)

PFAS are man-made chemicals that have been used in industry and consumer products worldwide since the 1950s. They have been used in non-stick cookware; water-repellent clothing; firefighting foams; and products that resist grease, water, and oil. The most commonly studied PFAS are perfluorooctanoic acid (PFOA) and perfluorooctane sulfonic acid (PFOS). Some, but not all, studies in humans with PFAS exposure have shown that certain PFAS may affect growth, learning, and behavior of children; lower a woman's chance of getting pregnant; interfere with the body's natural hormones; increase cholesterol levels; affect the immune system; and increase the risk of cancer. Scientists are still learning about the health effects of exposures to mixtures of PFAS.

## PHOSGENE

Phosgene is a colorless, nonflammable gas that has the odor of freshly cut hay. Exposure to low levels can irritate the eyes and throat, and higher levels can cause the lungs to swell, making it difficult to breathe; this can happen quickly or might not be noticed until the next day. Exposure to high levels can result in severe damage to the lungs that might lead to death.

## PLUTONIUM

Plutonium is a silvery-white radioactive metal. The most common plutonium isotopes are plutonium-238 and plutonium-239. The half-life, the time it takes for half of the plutonium to undergo radioactive decay and change forms, is 87.7 years for plutonium-238 and 24,100 years for plutonium-239. The main health effect from exposure to plutonium is cancer, which may occur years after exposure. The types of cancers most likely to develop are cancers of the lungs, bones, and liver.

## POLYCHLORINATED BIPHENYLS (PCBS)

Polychlorinated biphenyls have been used as coolants and lubricants in transformers, capacitors, and other electrical equipment because they don't burn easily and are good insulators. The manufacture of PCBs was stopped in the United States in 1977, because of evidence they build up in the environment and damage health. The most commonly observed health effects are skin conditions, and studies in exposed workers have shown changes in blood and urine that may indicate liver damage. Babies born to women who ate PCB-contaminated fish showed abnormal responses in tests of infant behavior, for example, problems with motor skills and a decrease in short-term memory, some of which lasted for several years. Other studies suggest that the immune system was affected in children born to and nursed by mothers exposed to increased levels of PCBs. PCBs have been classified as probably carcinogenic by the EPA and carcinogenic by the International Agency for Research on Cancer.

## RADON

Radon is an odorless, colorless, naturally occurring radioactive gas formed from the breakdown of uranium and thorium. Radon progeny is the term given to those radioactive atoms with short half-lives into which radon quickly decays. Radon and radon progeny are normally found at higher levels indoors. When radon or radon progeny undergo radioactive decay, some of the decays expel high-energy alpha particles, which are the main source of health concerns. Many scientists believe long-term exposure to elevated levels of radon progeny in air increases the chance of developing lung cancer. Smaller lungs and faster breathing rates may result in higher radiation doses to the lungs of children relative to adults.


## STRONTIUM

Strontium is a naturally occurring element found in rocks, soil, dust, coal, and oil. It can also exist as several radioactive isotopes; the most common is strontium-90, which has a half-life of 29 years. Breathing or ingesting low levels of radioactive strontium have not been shown to affect health. High levels of radioactive strontium can damage bone marrow and cause anemia, and prevent the blood from clotting properly. Exposure to high levels of radioactive strontium may cause cancer.


## TRICHLOROETHYLENE (TCE)

Trichloroethylene is a colorless, volatile liquid with a sweet odor. One of its main uses is as a solvent to remove grease from metal parts. People who are overexposed to moderate amounts of trichloroethylene may experience headaches, dizziness, and sleepiness; large amounts of trichloroethylene may cause coma and death. Other effects seen in people exposed to high levels of trichloroethylene include evidence of nervous system effects related to hearing, seeing, and balance; changes in the rhythm of the heartbeat; liver damage; and evidence of kidney damage. Exposure to trichloroethylene may cause scleroderma (a systemic autoimmune disease), and some men may experience a decrease in sex drive, sperm quality, and reproductive hormone levels. There is strong evidence that trichloroethylene can cause kidney cancer and some evidence that it causes liver cancer and malignant lymphoma, a blood cancer.


## URANIUM

Uranium is a naturally occurring radioactive element that is almost as hard as steel and much denser than lead. Natural uranium is used to make enriched uranium; depleted uranium is the leftover product. Enriched uranium is used to make fuel for nuclear power plants. Uses for depleted uranium include as a counterbalance on helicopter rotors and a component of munitions to help them penetrate armored vehicles. Natural uranium and depleted uranium have an identical chemical effect on the body. Kidney damage has been seen in humans and animals after inhaling or ingesting uranium compounds. Health effects of natural and depleted uranium are due to chemical effects and not radiation. Some studies suggest that exposure to depleted uranium can increase the frequency of birth defects, but they do not allow valid conclusions.

## WHITE PHOSPHORUS

White phosphorus is a colorless, white or yellow, waxy solid with a garlic-like odor that reacts rapidly with oxygen, easily catching fire. White phosphorus is used by the military in various types of ammunition and to produce smoke for concealing troop movements and identifying targets. Breathing white phosphorus for short periods may cause coughing and irritation of the throat and lungs; breathing it for long periods may cause poor wound healing of the mouth and breakdown of the jaw bone. Consuming small amounts of white phosphorus can cause liver, heart, or kidney damage; stomach cramps; or death. Skin contact with burning white phosphorus may cause liver, heart, and kidney damage.

*Source: Agency for Toxic Substances and Disease Registry.*

# NOTES

## INTRODUCTION

1. Ui Jun (ed.), *Industrial Pollution in Japan* (Tokyo: United Nations University Press, 1992), 183.

2. "Ensure a Safe Environment through Clearing the Land from UXO and Educating the Population about Risks," *United Nations in Laos PDR*, http://www.la.one .un.org/sdgs/sdg-18-lives-safe-from-uxo (accessed June 14, 2019).

3. For an introduction to military contamination in the former Soviet Union, see Paul Josephson, "War on Nature as Part of the Cold War: The Strategic and Ideological Roots of Environmental Degradation in the Soviet Union," in *Environmental Histories of the Cold War*, ed. J. R. McNeill and Corinna R. Unger (New York: Cambridge University Press, 2010), 44–46.

4. Rob Evans, "Military Scientists Tested Mustard Gas on Indians," *Guardian*, September 1, 2007, https://www.theguardian.com/uk/2007/sep/01/india.military (accessed June 14, 2019); quoted in Tilman A. Ruff, "The Humanitarian Impact and Implications of Nuclear Test Explosions in the Pacific Region," *International Review of the Red Cross* 97, no. 899 (2015): 778–79, doi:10.1017/S1816383116000163; Andrew Buncombe, "China's Secret Nuclear Tests Leave Legacy of Cancer and Deformity," *Independent*, October 5, 1998, https://www.independent.co.uk/news /chinas-secret-nuclear-tests-leave-legacy-of-cancer-and-deformity-1176260.html (accessed June 14, 2019); "China Nuclear Tests Prompt Uighur Campaign," *Japan Times*, August 9, 2012, https://www.japantimes.co.jp/news/2012/08/09/national /china-nuclear-tests-prompt-uighur-campaign/ (accessed June 14, 2019).

5. "United States Nuclear Tests: July 1945 through September 1992," *US Department of Energy, National Nuclear Security Administration*, September 2015, https://www.nnss.gov/docs/docs_LibraryPublications/DOE_NV-209_Rev16.pdf (accessed June 14, 2019); *Hiroshima in America*, 322–331.

6. "How the US Government Exposed Thousands of Americans to Lethal Bacteria to Test Biological Warfare," *Democracy Now*, July 13, 2005, https://www .democracynow.org/2005/7/13/how_the_u_s_government_exposed (accessed June 14, 2019).

## CHAPTER 1: JAPANESE WEAPONS OF MASS DESTRUCTION AND THE US COVER-UP

1. "History: Looking Back Helps Us to Look Forward," *Organization for the Prohibition of Chemical Weapons*, https://www.opcw.org/about-us/history (accessed June 14, 2019); Tanaka Yuki, "Poison Gas: The Story Japan Would Like to Forget," *Bulletin of the Atomic Scientists* (October 1988): 11.

2. Tanaka, "Poison Gas," 12; interview with Yamauchi Masayuki, February 22, 2019.

3. Chuugoku Shimbun, *Dokugasu no Shima: Okunoshima – Akumu no Kizuato* (Hiroshima: Chuugoku Shimbunsha, 1996), 10.

4. Interview with Yamauchi, February 22, 2019.

5. Chuugoku, *Dokugasu no Shima*, 13–14; interview with Yamauchi, February 22, 2019.

6. Chuugoku, *Dokugasu no Shima*, 22; "Kyuu nippongun no dokugasu heiki/ Watashi wa 'oni' ni sareta, moto youseikou," *Shimbun Akahata*, November 23, 2015, http://www.jcp.or.jp/akahata/aik15/2015-11-23/2015112303_01_0.html (accessed June 14, 2019); Chuugoku, *Dokugasu no Shima*, 13.

7. "Shouwa 48 nen no 'Kyuugun dokugasu dan nado no zenkoku chousa' foroappu chousa houkokusho," *Ministry of Environment*, November 2003, 1, https:// www.env.go.jp/chemi/report/h15-02/ (accessed June 14, 2019), hereafter referred to as "2003 Follow-up Survey"; Chuugoku, *Dokugasu no Shima,* 64.

8. "Okunoshima no dokugasu no rekishi no gaiyou," *Okunoshima kara heiwa to kankyou o kangaeru kai*, http://dokugas.server-shared.com/new_page_32.htm (accessed June 14, 2019).

9. Tanaka, "Poison Gas," 16; Office of the Chief Chemical Officer, "Intelligence Report on Japanese Chemical Warfare," General Headquarters, May 15, 1946, 128, 63, 144.

10. Chuugoku, *Dokugasu no Shima*, 83.

11. Tanaka, "Poison Gas," 15; Walter E. Grunden, "No Retaliation in Kind: Japanese Chemical Warfare Policy in World War II," in *One Hundred Years of Chemical Warfare: Research, Deployment, Consequences*, ed. Bretislav Friedrich, Dieter Hoffmann, Jürgen Renn, Florian Schmaltz, and Martin Wolf (Cham: Springer Nature, 2017), 262; Tanaka, "Poison Gas," 17.

12. Haruko Taya Cook and Theodore F. Cook, *Japan at War: An Oral History* (New York: New Press, 1992), 44; Office of the Chief Chemical Officer, "Intelligence Report on Japanese Chemical Warfare," 14; Sheldon H. Harris, *Factories*

*of Death: Japanese Biological Warfare, 1932–1945, and the American Cover-up* (London: Routledge, 1994), 73.

13. Grunden, "No Retaliation in Kind," 267. For the discovery of Japanese chemical weapons on Okinawa, see chapter 3.

14. Daqing Yang, "Documentary Evidence and Studies of Japanese War Crimes: An Interim Assessment," in *Researching Japanese War Crimes Records: Introductory Essays*, ed. Edward Drea, Greg Bradsher, Robert Hanyok, James Lide, Michael Petersen, and Daqing Yang (Washington, D.C.: Nazi War Crimes and Japanese Imperial Government Records Interagency Working Group, 2006), 38.

15. "Japan Tested Chemical Weapon on Aussie POW: New Evidence," *Japan Times*, July 27, 2004, https://www.japantimes.co.jp/news/2004/07/27/national/japan-tested-chemical-weapon-on-aussie-pow-new-evidence/ (accessed June 14, 2019).

16. Jonathan B. Tucker, *War of Nerves: Chemical Warfare from World War I to Al-Qaeda* (New York: Anchor Books, 2006), 102; William A. Buckingham, *Operation Ranch Hand: The Air Force and Herbicides in Southeast Asia, 1961–1971* (Washington, D.C.: Office of Air Force History, 1982), 4; Sheldon H. Harris, *Factories of Death: Japanese Biological Warfare, 1932–1945, and the American Cover-up* (London: Routledge, 1994), 156.

17. Chuugoku, *Dokugasu no Shima*, 66.

18. Chuugoku, *Dokugasu no Shima*, 66; interview with Yamauchi, February 22, 2019.

19. Tanaka, "Poison Gas," 18; interview with Yamauchi, February 22, 2019.

20. "2003 Follow-up Survey," 40.

21. "2003 Follow-up Survey," 40.

22. Interview with Yamauchi, February 22, 2019; Tanaka, "Poison Gas," 19.

23. Interview with Yamauchi, February 22, 2019.

24. Chuugoku, *Dokugasu no Shima*, 44–46, 51–54.

25. Inagaki Mihoko, "Musekinin sugiru, dokugasu higai taisaku," *Shuukan Kinyoubi*, August 18, 2017, 33.

26. "2003 Follow-up Survey," 18–20. In April 1951, four people died when an abandoned shell they had taken home to dismantle for scrap leaked an unspecified gas; then, in June 1954, another six workers were injured by mustard agent while attempting to take apart a munition. In September 1957, ten members of a trawler crew were injured by a leaking three-hundred-kilogram barrel of mustard agent. In the next twelve years, thirty-eight more people would be injured in the Choushi area.

27. "2003 Follow-up Survey," 24–25.

28. "2003 Follow-up Survey," 24.

29. Chuugoku, *Dokugasu no Shima*, 75–78.

30. "Poison Gas Shell Disposal Commences in Hokkaido," *Japan Times*, September 24, 2000, https://www.japantimes.co.jp/news/2000/09/24/national/poison-gas-shell-disposal-commences-in-hokkaido/#.XfPDJ3t7nIU (accessed June 14, 2019).

31. "2003 Follow-up Survey," 39; "Japan's Disarmament and Nonproliferation Policy," *Ministry of Foreign Affairs*, March 2006, 127, https://www.mofa.go.jp /policy/un/disarmament/policy/pamph0603/1-4.pdf (accessed June 14, 2019).

32. "OPCW Urged to Ensure Destruction of Chemical Weapons Abandoned by Japan in China," *XinhuaNet*, November 28, 2017, http://www.xinhuanet.com// english/2017-11/28/c_136785861.htm (accessed June 14, 2019); "WW2 Chemical Weapons Injure Two Boys," *China Daily,* July 28, 2004. In 1974, more than thirty workers were exposed on a pump boat on a river in Jiamusi City, Heilongjiang Province, when they pulled up a shell of suspected mustard agent. Five more were hurt by abandoned chemical ordnance in the same province in 1982 (Chuugoku, *Dokugasu no Shima*, 97–100). As China's economy prospered and previously un-developed areas became urbanized, the number of incidents increased. One of the most publicized events occurred in 2003, in Qiqihar, Heilongjiang Province, when five barrels of blister agent injured forty-three people and killed one. The follow-ing year, two boys were injured in Jilin Province, when they were exposed to other abandoned chemical weapons ("China Mustard Gas Crisis," *Reuters*, August 19, 2003, https://www.nytimes.com/2003/08/19/world/china-mustard-gas-crisis.html [accessed June 14, 2019]; "WW2 Chemical Weapons Injure Two Boys").

33. Mike Brombach, "Abandoned Chemical Weapons in China: The Unresolved Japanese Legacy," *Global Green USA*, May 2011, https://static1.squarespace .com/static/5548ed90e4b0b0a763d0e704/t/55548b35e4b031f59111a514 /1431604021283/publication-183-1.pdf (accessed June 14, 2019).

34. "Ceremony Marks Start of Destruction of Chemical Weapons Abandoned by Japan in China," *Organization for the Prohibition of Chemical Weapons*, Sep-tember 8, 2010, https://www.opcw.org/media-centre/news/2010/09/ceremony-marks -start-destruction-chemical-weapons-abandoned-japan-china (accessed June 14, 2019); "Chuugoku ni okeru ikikagakuheiki mondai ni kan suru dai 20 kai nichichuu kyoudou sagyou guru-pu kaigou," *Ministry of Foreign Affairs*, March 22, 2018, https:// www.mofa.go.jp/mofaj/a_o/c_m1/cn/page4_003860.html (accessed June 14, 2019).

35. Sheldon H. Harris, *Factories of Death: Japanese Biological Warfare, 1932–1945, and the American Cover-up* (London: Routledge, 1994), 57.

36. Harris, *Factories of Death*, 25–26. Anthrax triggers different symptoms de-pending on whether it is inhaled, swallowed, or injected. Gastrointestinal anthrax, for example, can cause victims to vomit blood, and injected anthrax causes the skin to develop blisters and abscesses; each form can be fatal. What attracted Japanese scientists to glanders was its ability to infect horses, the primary means to traffic troops and supplies throughout the region. Highly contagious, glanders attacks the animals' lungs and respiratory tract; it can also infect humans. Bubonic plague, aka the Black Death, wiped out approximately 60 percent of the population of Europe in the fourteenth century. The disease is spread by infected fleas, and one of the chief symptoms is swollen lymph nodes, which circulate bacteria through the body, killing skin and organs, and turning them black.

37. Harris, *Factories of Death*, 34; Daniel Barenblatt, *A Plague upon Humanity* (New York: HarperCollins, 2004), 49.

38. Barenblatt, *A Plague upon Humanity*, 42, 46.

39. William H. Cunliffe, *Select Documents on Japanese War Crimes and Japanese Biological Warfare, 1934–2006* (Washington, D.C.: National Archives and Records Administration, 2007), 32; Barenblatt, *A Plague upon Humanity*, 72–73; Harris, *Factories of Death*, 55.

40. Cunliffe, *Select Documents on Japanese War Crimes and Japanese Biological Warfare*, 32; Harris, *Factories of Death*, 61.

41. Peter Williams and David Wallace, *Unit 731: The Japanese Army's Secret of Secrets* (London: Hodder and Stoughton, 1989), 41–42; Barenblatt, *A Plague upon Humanity*, 72.

42. Tanaka, "Poison Gas," 14; Williams and Wallace, *Unit 731*, 46.

43. Quoted in Russell Working, "The Trial of Unit 731," *Japan Times*, June 5, 2001, https://www.japantimes.co.jp/opinion/2001/06/05/commentary/world-com mentary/the-trial-of-unit-731 (accessed June 14, 2019).

44. Melody Zaccheus, "WWII S'pore Used as Base to Spread Disease," *Straits Times*, November 13, 2017, https://www.straitstimes.com/singapore/wwii -spore-used-as-base-to-spread-disease (accessed June 14, 2019); Justin McCurry, "Japanese Veteran Admits Vivisection Tests on POWs," *Guardian*, November 27, 2006, https://www.theguardian.com/world/2006/nov/27/secondworldwar.japan (accessed June 14, 2019); Justin McCurry, "Japan Revisits Its Darkest Moments Where American POWs Became Human Experiments," *Guardian*, August 13, 2015, https:// www.theguardian.com/world/2015/aug/13/japan-revisits-its-darkest-moments-where -american-pows-became-human-experiments (accessed June 14, 2019).

45. Williams and Wallace, *Unit 731*, 64–65.

46. Harris, *Factories of Death*, 77.

47. Cunliffe, *Select Documents on Japanese War Crimes and Japanese Biological Warfare*, 12; Barenblatt, *A Plague upon Humanity*, 143–44.

48. Williams and Wallace, *Unit 731*, 68–70.

49. Barenblatt, *A Plague upon Humanity*, 77, 157; Harris, *Factories of Death*, 100.

50. Harris, *Factories of Death*, 104.

51. Barenblatt, *A Plague upon Humanity*, 173; Nicholas D. Kristof, "Unmasking Horror: Japan Confronting Gruesome War Atrocity," *New York Times*, March 17, 1995, 1.

52. Tracy Dahl, "Japan's Germ Warriors," *Washington Post*, May 26, 1983, https:// www.washingtonpost.com/archive/politics/1983/05/26/japans-germ-warriors /a0149d21-ba27-460e-a807-d3db942ba507/ (accessed June 14, 2019); Cunliffe, *Select Documents on Japanese War Crimes and Japanese Biological Warfare*, 47; Kristof, "Unmasking Horror," 1.

53. Chuugoku, *Dokugasu no Shima*, 108; Barenblatt, *A Plague upon Humanity*, 172.

54. Cunliffe, *Select Documents on Japanese War Crimes and Japanese Biological Warfare*, 46–47.

55. Cunliffe, *Select Documents on Japanese War Crimes and Japanese Biological Warfare*, 48.

56. Christopher Reed, "The United States and the Japanese Mengele: Payoffs and Amnesty for Unit 731," *Asia-Pacific Journal* 4, no, 8 (August 14, 2006): 1–6, https://apjjf.org/-Christopher-Reed/2177/article.html (accessed June 14, 2019).

57. NARA staff, "Japanese War Crimes Records at the National Archives: Research Starting Points," in *Researching Japanese War Crimes Records: Introductory Essays*, ed. Edward Drea, Greg Bradsher, Robert Hanyok, James Lide, Michael Petersen, and Daqing Yang (Washington, D.C.: Nazi War Crimes and Japanese Imperial Government Records Interagency Working Group, 2006), 106.

58. Quoted in Reed, "The United States and the Japanese Mengele."

59. Harris, *Factories of Death*, 213.

60. See, for example, "Rikugun Noborito Kenkyujo no shinjitsu [The Truth about the Army Noborito Research Institute]—Intelligence in Recent Public Literature," *Central Intelligence Agency*, https://www.cia.gov/library/center-for -the-study-of-intelligence/csi-publications/csi-studies/studies/vol46no4/article11 .html (accessed June 14, 2019); "Report of the International Scientific Commission for the Investigation of the Facts Concerning Bacterial Warfare in Korea and China," *Credo*, 1952, 63, https://credo.library.umass.edu/view/pageturn /mums312-b139-i045/#page/2/mode/1up (accessed June 14, 2019); "Exploitation of Communist BW charges," Psychological Strategy Board memorandum, *Central Intelligence Agency*, July 7, 1953, https://www.cia.gov/library/readingroom/docs /CIA-RDP80R01731R003300190004-6.pdf (accessed June 14, 2019).

61. Kristof, "Unmasking Horror," 1; Harris, *Factories of Death*, 324; Barenblatt, *A Plague upon Humanity*, 234.

62. Jonathan Watts, "Japan Guilty of Germ Warfare against Thousands of Chinese," *Guardian*, August 28, 2002, https://www.theguardian.com/world/2002 /aug/28/artsandhumanities.japan (accessed June 14, 2019).

## CHAPTER 2: NUCLEAR WARFARE IN JAPAN AND THE MARSHALL ISLANDS

1. Andrew J. Rotter, *Hiroshima: The World's Bomb* (New York: Oxford University Press, 2008), 65–66, 76–78. For an in-depth (English) exploration of Japan's attempts to build a bomb, see also John W. Dower, "'NI' and 'F': Japan's Wartime Atomic Bomb Research," in *Japan in War and Peace: Essays on History, Culture, and Race* (London: HarperCollins, 1995), 55–100.

2. Richard Rhodes, *The Making of the Atomic Bomb: The 25th Anniversary Edition* (New York: Simon and Schuster Paperbacks, 2012), Loc. 12687.

3. Andrew J. Rotter, *Hiroshima: The World's Bomb* (New York: Oxford University Press, 2008), 115; Jonathan M. Weisgall, *Operation Crossroads: The Atomic Tests at Bikini Atoll* (Annapolis, Md.: Naval Institute Press, 1994), 138–40. Just how little US scientists understood the new field of nuclear physics was evident in the summer of 1942. when they temporarily halted work after wondering whether a

nuclear explosion might trigger a catastrophic chain reaction in the nitrogen in the atmosphere. After calculating that the risk was (by their standards) an acceptable three in 1 million, they went back to work (Rotter, *Hiroshima*, 121).

4. "United States Nuclear Tests: July 1945 through September 1992," *US Department of Energy, National Nuclear Security Administration*, September 2015, 25, https://www.nnss.gov/docs/docs_LibraryPublications/DOE_NV-209_Rev16.pdf (accessed June 14, 2019); Robert Jay Lifton and Greg Mitchell, *Hiroshima in America: Fifty Years of Denial* (New York: G. P. Putnam's Sons, 1995), 155; "The Manhattan Project: An Interactive History," *US Department of Energy*, https://www.osti.gov/opennet/manhattan-project-history/Events/1945/trinity_safety.htm (accessed June 14, 2019).

5. Quoted in Rhodes, *The Making of the Atomic Bomb*, Loc. 14120, 14135.

6. Lifton and Mitchell, *Hiroshima in America*, 231, 23.

7. "The Atomic Bombings of Hiroshima and Nagasaki," Manhattan Engineer District of the United States Army, June 29, 1946, Chapter 3.

8. Manhattan, "Atomic Bombings," Chapter 17.

9. "Frequently Asked Questions about the Atomic-Bomb Survivor Research Program," *Radiation Effects Research Foundation*, https://www.rerf.or.jp/en/faq/ (accessed June 12, 2019).

10. Hiroshima Peace Memorial Museum, *The Spirit of Hiroshima* (Hiroshima: Hiroshima Peace Memorial Museum, 1999), 71; Masuda Sakiko, "'Black Rain' Radiation from 1945 Hiroshima A-Bombing Can Still Be Detected on Survivor's Shirt: Study," *Japan Times*, June 25, 2018, https://www.japantimes.co.jp/news/2018/08/06/national/history/black-rain-radiation-1945-hiroshima-bombing-can-still-detected-survivors-shirt-study/#.XfdvBnt7nIU (accessed June 12, 2019).

11. Data from Hiroshima Peace Memorial Museum, February, 2019.

12. Oliver Stone and Peter Kuznick, *The Untold History of the United States* (New York: Simon and Schuster, 2012), 168–69; Lifton and Mitchell, *Hiroshima in America*, 163; Susan Southard, *Nagasaki: Life after Nuclear War* (New York: Penguin, 2016), 84.

13. Data from Hiroshima Peace Memorial Museum, February, 2019; Southard, *Nagasaki*, 177, 141.

14. George Weller, *First into Nagasaki: The Censored Eyewitness Dispatches on Post-Atomic Japan and Its Prisoners of War* (New York: Crown, 2006), 134, 3.

15. Weller, *First into Nagasaki*, 254, 258.

16. Weller, *First into Nagasaki*, 270.

17. "Statement by the President Announcing the Use of the A-Bomb at Hiroshima," *Harry S. Truman Library and Museum*, August 6, 1945, https://www.trumanlibrary.org/publicpapers/index.php?pid=100 (accessed June 12, 2019); quoted in Southard, *Nagasaki*, 66.

18. Southard, *Nagasaki*, 112–13.

19. "The Atomic Bombings of Hiroshima and Nagasaki."

20. Rotter, *Hiroshima*, 221.

21. Weller, *First into Nagasaki*, 4, 303–10.

22. Monica Braw, *The Atomic Bomb Suppressed: American Censorship in Occupied Japan* (New York: M. E. Sharpe, 1991), 93, 99, 101.

23. Sheldon H. Harris, *Factories of Death: Japanese Biological Warfare, 1932–1945, and the American Cover-up* (London: Routledge, 1994), 324.

24. For example, see Southard, *Nagasaki*, 221–224.

25. Henry L. Stimson, "The Decision to Use the Atomic Bomb," Harper's Magazine, February 1947, 97-107.

26. Rotter, *Hiroshima*, 176; Lifton and Mitchell, *Hiroshima in America*, 273, 240; Rotter, *Hiroshima*, 162.

27. "Text of Hirohito's Radio Rescript," *The New York Times*, August 15, 1945, 3.

28. Lifton and Mitchell, *Hiroshima in America*, 327.

29. Jane Dibblin, *Day of Two Suns: US Nuclear Testing and the Pacific Islanders* (New York: New Amsterdam Books, 1990), 17.

30. Quoted in Donald F. McHenry, *Micronesia: Trust Betrayed* (Washington, D.C.: Carnegie Endowment for International Peace, 1975), 87.

31. Weisgall, *Operation Crossroads*, 107.

32. Weisgall, *Operation Crossroads*, 120.

33. Weisgall, *Operation Crossroads*, 199.

34. *Project Crossroads*, US Department of Energy, Albuquerque Operations Office, 1946. Available to watch: https://youtu.be/2HkLZekOZLU

35. Weisgall, *Operation Crossroads*, 223.

36. Weisgall, *Operation Crossroads*, 230, 238.

37. Quoted in Weisgall, *Operation Crossroads*, 266.

38. Weisgall, *Operation Crossroads*, 275.

39. Braw, *The Atomic Bomb Suppressed*, 116.

40. Ricardo M. Gonzalez and Mark D. Merlin, "Environmental Impacts of Nuclear Testing in Remote Oceania, 1946–1996," in *Environmental Histories of the Cold War*, ed. J. R. McNeill and Corinna R. Unger (New York: Cambridge University Press, 2010), 192, 177; Weisgall, *Operation Crossroads*, 302.

41. Jane Dibblin, *Day of Two Suns*, 24–27, 58.

42. Dibblin, *Day of Two Suns*, 29.

43. Matashichi Oishi, *The Day the Sun Rose in the West: Bikini, the Lucky Dragon, and I* (Honolulu: University of Hawai'i Press, 2011), 122.

44. Oishi, *The Day the Sun Rose in the West*, 19–20.

45. Oishi, *The Day the Sun Rose in the West*, 42, 27, 35; Weisgall, *Operation Crossroads*, 305.

46. Oishi, *The Day the Sun Rose in the West*, 126.

47. Oishi, *The Day the Sun Rose in the West*, 99, 142.

48. Popular culture also tapped into this zeitgeist with the release of the movie *Gojira* in November 1954. Far bleaker than the sanitized US version and Japanese sequels, its images played off Hiroshima, Nagasaki, and Bikini: a fishing ship struck

by a *pikadon*, frightened children checked by Geiger counters, and cities razed to seas of flame. Its final scene featured a call for the end of nuclear testing.

49. "A Reactor for Japan," *Washington Post*, September 23, 1954, 18.

50. Gonzalez and Merlin, "Environmental Impacts of Nuclear Testing in Remote Oceania, 1946–1996," 194.

51. Gonzalez and Merlin, "Environmental Impacts of Nuclear Testing in Remote Oceania, 1946–1996," 178.

52. Weisgall, *Operation Crossroads*, 304; quoted in Giff Johnson, "Study Calls Marshall Islands' Cancer Rate Extreme," *Pacific Islands Report*, March 22, 1999, http://www.pireport.org/articles/1999/03/22/study-calls-marshall-islands-cancer-rate-extreme (accessed June 14, 2019); Dibblin, *Day of Two Suns*, 36.

53. Dibblin, *Day of Two Suns*, 36, 40.

54. "Marshall Islands," *Atomic Heritage Foundation*, https://www.atomicheritage.org/location/marshall-islands (accessed June 14, 2019).

55. Tilman A. Ruff, "The Humanitarian Impact and Implications of Nuclear Test Explosions in the Pacific Region," *International Review of the Red Cross* 97, no. 899 (2015): 795.

56. Dave Philipps, "Troops Who Cleaned Up Radioactive Islands Can't Get Medical Care," *New York Times*, January 28, 2017, https://www.nytimes.com/2017/01/28/us/troops-radioactive-islands-medical-care.html (accessed June 14, 2019).

57. Interview with Robert Celestial, March 13, 2019.

58. Debra Killalea, "Marshall Islands: Concrete Dome Holding Nuclear Waste Could Leak," *News.com.au*, November 27, 2017, https://www.news.com.au/technology/environment/marshall-islands-concrete-dome-holding-nuclear-waste-could-leak/news-story (accessed June 14, 2019).

59. Philipps, "Troops Who Cleaned Up Radioactive Islands Can't Get Medical Care."

60. Coleen Jose, Kim Wall, and Jan Hendrik Hinzel, "This Dome in the Pacific Houses Tons of Radioactive Waste—and It's Leaking," *Guardian*, July 3, 2015, https://www.theguardian.com/world/2015/jul/03/runit-dome-pacific-radioactive-waste (accessed June 14, 2019).

61. "The Legacy of US Nuclear Testing and Radiation Exposure in the Marshall Islands," *US Embassy in the Republic of The Marshall Islands*, https://mh.usembassy.gov/the-legacy-of-u-s-nuclear-testing-and-radiation-exposure-in-the-marshall-islands/ (accessed June 14, 2019).

62. "Projected Costs of US Nuclear Forces, 2019 to 2028," *Congressional Budget Office*, January 24, 2019, https://www.cbo.gov/publication/54914 (accessed June 14, 2019).

63. Autumn S. Bordner, Danielle A. Crosswella, Ainsley O. Katza, Jill T. Shaha, Catherine R. Zhanga, Ivana Nikolic-Hughes, Emlyn W. Hughesa, and Malvin A. Rudermand, "Measurement of Background Gamma Radiation in the Northern

Marshall Islands," *PNAS* 113, no. 25 (June 21, 2016): 6,833–38, first published June 6, 2016, https://doi.org/10.1073/pnas.1605535113.

64. "The Legacy of US Nuclear Testing and Radiation Exposure in the Marshall Islands."

65. Anthony Robbins, Arjun Makhijani, and Katherine Yih, *Radioactive Heaven and Earth: The Health and Environmental Effects of Nuclear Weapons Testing in, on, and above the Earth*, Report of the IPPNW International Commission to Investigate the Health and Environmental Effects of Nuclear Weapons Production and Institute for Energy and Environmental Research (New York and London: Apex Press/Zed Books, 1991), 34–40; Frederick Warner and René J. C. Kirchmann (eds.), *Nuclear Test Explosions: Environmental and Human Impacts*, Scientific Committee on Problems of the Environment (SCOPE) of the International Council for Science, SCOPE report, no. 59 (New York: John Wiley & Sons, 1999), 220–21.

## CHAPTER 3: OKINAWA: "THE JUNK HEAP OF THE PACIFIC"

1. Clarence J. Glacken, *The Great Loochoo: A Study of Okinawan Village Life* (Berkeley: University of California Press, 1955), 32–37.

2. Ryukyu Shimpo, *Descent into Hell: Civilian Memories of the Battle of Okinawa* (Portland, Maine: MerwinAsia, 2014), xviii; Ota Masahide, *Essays on Okinawa Problems* (Gushikawa: Yui Shuppan Company, 2000), 58.

3. Ted Tsukiyama, "The Battle of Okinawa," *The Hawai'i Nisei Story: Americans of Japanese Ancestry during WW2, University of Hawaii*, http://nisei.hawaii .edu/object/io_1149316185200.html (accessed June 12, 2019); Ryukyu Shimpo, *Descent into Hell*, 184; "Fuhatsudan de shibou/Sengo 710 nin/Jiko 1087 ken," *Okinawa Times*, June 17, 2009, 1; "Fuhatsudan zero ni ato nannen?" *Okinawa Times*, June 23, 2017, http://www.okinawatimes.co.jp/articles/-/104905 (accessed June 12, 2019). The most horrific incident occurred in August 1948, when a boat carrying rockets from Iejima, a small island off Okinawa's northwest coast, exploded, killing or injuring 178 people. The subsequent US investigation blamed the accident on careless safety procedures. In March 1974, buried ordnance blew up during construction work near a kindergarten in Oroku, Naha City, killing four people, one of whom was a three-year old girl, and wounding thirty-four. More recently, in January 2009, a 250-kilogram US bomb exploded in Itoman when it was struck by a digging machine; the blast injured two and shattered one hundred windows at a nearby nursing home ("1948 nen no beigun LCT bakuhatsu de atarashii shiryou," *Okinawa Times*, February 15, 2017, https://www.okinawatimes.co.jp/articles/-/84288 [accessed June 12, 2019]; "Monument Built to Honor Victims of Ordinance Explosion at St. Matthew Kindergarten," *Ryukyu Shimpo*, March 2, 2015, http://english .ryukyushimpo.jp/2015/03/09/17368/ [accessed June 12, 2019]; "Kako ni mo tabitabi jiko hassei," *Ryukyu Asahi Housou*, aired January 14, 2009, https://www.qab.co.jp /news/20090114432.html [accessed June 12, 2019]).

4. "Amphibious Corps Action Report—Ryukyus Operation Phases I and II (Okinawa)," US Army, July 1, 1945, chapter VII.

5. In addition, the US military made widespread use of M15 grenades containing white phosphorus, a toxic metal that combusts on contact with air. Each grenade contained 425 grams of the substance, which spread over a seventeen-meter area; anyone unlucky enough to ingest even a few milligrams of this shrapnel faced organ damage. Just like other UXO, white phosphorus munitions continue to cause problems on Okinawa today. In June 2001, a bulldozer driver in Nishihara was injured by an exploding rocket; in October 2010, a white phosphorus shell started smoking in the school grounds of an elementary school in Yaese Town, where construction work was taking place (Ryukyu Shimpo, *Descent into Hell*, 185; "Shogakkou de ourindan mitsukaru," *Ryukyu Asahi Housou*, aired October 29, 2010, https://www.qab.co.jp/news/2010102922375.html [accessed June 12, 2019]).

6. "Kyuugun dokugasu dan nado no zenkoku chousa no foro-appu chousa ni tsuite—Okinawa ken," *Ministry of the Environment*, https://www.env.go.jp/chemi/report/h15-02/47-1itoman.pdf (accessed June 12, 2019). The Japanese government's 2003 chemical weapon follow-up survey also noted two other discoveries of suspected hydrogen cyanide grenades on Okinawa; they were encased in concrete and dumped at sea.

7. "Amphibious Corps Action Report—Ryukyus Operation Phases I and II (Okinawa)," US Army, chapter VII.

8. "Okinawa: Forgotten Island," *Time*, November 28, 1949, http://content.time.com/time/magazine/article/0,9171,856392,00.html (accessed June 12, 2019).

9. "Big Picture: Okinawa: Keystone of the Pacific," *National Archives and Records Administration, Department of Defense*, n.d.

10. Okinawa ken kyouiku iinkai (eds.), *Okinawa Ken Shi* (Naha: Okinawa ken kyouiku iinkai, 1974), chapter 10.

11. Herbert P. Bix, *Hirohito and the Making of Modern Japan* (New York: HarperCollins, 2000), 627.

12. "Report of a Special Subcommittee of the Armed Services Committee, House of Representatives Following an Inspection Tour October 14 to November 23, 1955" (aka, "The Price Report"), Washington, D.C., 1955.

13. Ahagon Shoko, *The Island Where People Live*, trans. C. Harold Rickard (Hong Kong: Christian Conference of Asia Communications, 1989), 100–101; *Scoop document: Okinawa to kaku*, Nippon Housou Kyoukai, aired September 10, 2017.

14. The best English-language exploration of the Okinawan migrant experience during this period is Steve Rabson, *The Okinawan Diaspora in Japan: Crossing the Borders Within* (Honolulu: University of Hawai'i Press, 2012), and Yoshida Kensei, *Democracy Betrayed: Okinawa under US Occupation* (Bellingham: Western Washington University, 2001), 69.

15. Nicholas Evan Sarantakes, *Keystone: The American Occupation of Okinawa and US–Japanese Relations* (College Station: Texas A&M University Press, 2000), 97; Roger B. Jeans, *The CIA and Third Force Movements in China during the Early Cold War: The Great American Dream* (Lanham, Md.: Lexington, 2018), 126.

16. Yara Tomohiro, "Naze beigun kichi wa okinawa ni shuuchuu shiteiru no ka," *New Diplomacy Initiative*, June 16, 2016, http://www.nd-initiative.org/research /5906/ (accessed June 12, 2019); quoted in Sarantakes, *Keystone*, 110.

17. The incident and its aftermath were reported in Okinawan newspapers, for example, "Chuushokuchuu no kyoushitsu ni sairui gasu?," *Okinawa Times*, March 12, 1965, 7 and "Sairuidan o tsukatta," *Okinawa Times*, March 13, 1965, 7. Leaks of CS gas continued during the Vietnam War, with two incidents in 1972, at Chibana Army Ammunition Depot, injuring several Americans and Okinawan base workers. Children again fell victim in January 1973, when CS gas seeped into Yomitan High School as students were in class, sickening dozens ("Beigun kichi kankyou karute: Kadena danyakukochiku," *Okinawa Prefecture*, March 2017, 20–11, https://www .pref.okinawa.lg.jp/site/kankyo/seisaku/documents/20.pdf [accessed June 12, 2019]).

18. Thomas R. H Havens, *Fire across the Sea: The Vietnam War and Japan, 1965–1975* (Princeton, N.J.: Princeton University Press, 1987), 87–88.

19. Quoted in Havens, *Fire across the Sea*, 85.

20. In June 1965, a trailer dropped by parachute crushed an eleven-year-old girl to death, and in November 1968, a bomb-laden B-52 crashed at Kadena Air Base and exploded, killing its crew, injuring sixteen Okinawans, and damaging more than three hundred buildings; if the plane had traveled several hundred more meters, it would have crashed into Chibana Ammunition Depot's stockpile of nuclear and chemical weapons. For a comprehensive, English-language list of US military crimes and accidents (1948–1995), see "List of Main Crimes Committed and Incidents Concerning the US Military on Okinawa," *Okinawa Peace Network of Los Angeles*, http://www.uchinanchu.org/history/list_of_crimes.html (accessed June 12, 2019). (The list was originally compiled by the *Okinawa Times*.)

21. In January 1968, a pipeline on MCAS Futenma broke, swamping 165,000m² of fields and rice paddies, contaminating the water source for 280 households. Another large spill of waste oil poured from Camp Kuwae in January 1973, contaminating the nearby coastline. Numerous large leaks of fuel also occurred in Naha, notably one in June 1974, that blocked the main road to the airport, causing chaos ("Beigun kichi kankyou karute: Futenma hikoujou," *Okinawa Prefecture*, March 2017, 49-15, http://www.pref.okinawa.jp/site/kankyo/seisaku/documents/49.pdf [accessed June 12, 2019]; "Beigun kichi kankyou karute: Camp kuwae," *Okinawa Prefecture*, March 2017, 41-11, http://www.pref.okinawa.jp/site/kankyo/seisaku/docu ments/41.pdf [accessed June 12, 2019]; "Beigun kichi kankyou karute: Rikugun choyushisetsu," *Okinawa Prefecture*, March 2017, 74-10, http://www.pref.okinawa .jp/site/kankyo/seisaku/documents/74.pdf [accessed June 12, 2019]).

22. The health dangers of fuel lie in two areas: the makeup of the fuel itself and the chemicals added to improve its performance. Fuel is comprised of many components, several of which, particularly benzene and naphthalene, are toxic. Exposure to benzene damages the body's ability to produce blood cells, and it can harm bone marrow and cause leukemia; naphthalene has been linked to damage to the central nervous system, kidneys, and liver. In addition, the military adds other

chemicals to inhibit corrosion and prevent ice from forming at the high altitudes at which aircraft fly. These substances have been linked to such health problems as damage to bone marrow and the male reproductive system.

23. "Okinawa Governor Raises Fuel Pipeline with USAGO," electronic telegram, *US Consulate Naha*, September 18 1974, https://search.wikileaks.org/plusd /cables/1974NAHA00397_b.html (accessed June 12, 2019).

24. "Operational Report—Lessons Learned, Headquarters, 2d Logistical Command, Period Ending 31 October 1969," US Army, January 28, 1970.

25. "15 nen mae no chatan doramukan/Watashi ga umeta," *Ryukyu Asahi Housou*, aired February 5, 2016, https://www.qab.co.jp/news/2016020577259.html (accessed June 12, 2019).

26. "Karehazai o abita shima: Betonamu to okinawa, moto beigunjin no shougen," *Ryukyu Asahi Housou*, aired May 15, 2012.

27. In February 1973, another spill of anticorrosion compound at Makiminato Service Area had forced 180 workers to flee their building, suffering pain in their eyes and coughing. "Beigun kichi kankyou karute: Makiminato hokyuuchiku," *Okinawa Prefecture*, March 2017, 54-13, https://www.pref.okinawa.jp/site/kankyo/seisaku/ documents/54.pdf (accessed June 12, 2019); "Okinawa Pollution Flap on Army Leak of Hexavalent Chromium," *US Consulate Naha*, August 25, 1975, https://wikileaks .org/plusd/cables/1975NAHA00432_b.html (accessed June 12, 2019).

28. "Base Pollution Incident on Okinawa," electronic telegram, US Consulate Naha, September 30 1975.

29. Robert S. Norris, William M. Arkin, and William Burr, "Where They Were," *Bulletin of the Atomic Scientists*, November/December 1999, 29; "Operation Red Hat: Men and a Mission," Department of Defense, 1971.

30. "Report of a Special Subcommittee of the Armed Services Committee, House of Representatives, Following an Inspection Tour October 14 to November 23, 1955."

31. "History of the Custody and Deployment of Nuclear Weapons: July 1945 through September 1977, Appendix B," *Office of the Secretary of Defense*, 1978, https://nsarchive2.gwu.edu/news/19991020/04-49.htm (accessed June 12, 2019). Not all the nuclear weapons stored on the island were slated for use by Okinawa-based forces. In the prelude to nuclear war, the USAF planned to fly some of the bombs to its airfields on mainland Japan in a move it dubbed "Operation High Gear." At these bases, the weapons would then be loaded onto USAF aircraft to attack targets in China and the Soviet Union. Despite the Japanese government's long-standing nonnuclear stance, the US military decided it not necessary to secure Japanese government approval before it launched the operation.

32. "First Atom Cannon Blast Closes Okinawa School," *Chicago Tribune*, October 28, 1955, 9; Abe Takashi, "Yume no kuni to kaku no shima," *Okinawa Times*, April 23, 2013, 1.

33. Daniel Ellsberg, *The Doomsday Machine: Confessions of a Nuclear War Planner* (New York: Bloomsbury, 2017), 48.

34. Jon Mitchell, "'Seconds Away from Midnight': US Nuclear Missile Pioneers on Okinawa Break Fifty-Year Silence on a Hidden Nuclear Crisis of 1962," *Asia-Pacific Journal* 10, issue 30, no. 1 (July 23, 2012): 1–11, https://apjjf.org/-Jon-Mitch ell/3800/article.pdf (accessed June 12, 2019). During the 1950s and 1960s, the United States considered the use of its nuclear weapons in Vietnam. In 1954, US government advisers debated loaning atomic bombs to the French during the battle of Dien Bien Phu in a last-ditch attempt to avoid surrendering their colony. Fourteen years later, as US troops faced similarly overwhelming forces at Khe Sanh, top commanders on Okinawa drafted a plan to drop nuclear weapons there; the operation was rejected by President Lyndon B. Johnson (Murrey Marder, "When Ike Was Asked to Nuke Vietnam," *Washington Post*, August 22, 1982, https://www .washingtonpost.com/archive/opinions/1982/08/22/when-ike-was-asked-to-nuke -vietnam/305c4152-202e-4303-9bdc-3424f4f7376b/ [accessed June 12, 2019]; David E. Sanger, "US General Considered Nuclear Response in Vietnam War, Cables Show," *New York Times*, October 6, 2018, https://www.nytimes.com /2018/10/06/world/asia/vietnam-war-nuclear-weapons.html [accessed June 12, 2019]).

35. Zaha Yukiyo, "Former US Soldier Details Account of 1959 Naha Accidental Nuke Firing Fatal to Fellow Soldiers," *Ryukyu Shimpo*, October 26, 2017, http://english.ryukyushimpo.jp/2017/11/01/27960/ (accessed June 12, 2019); Zaha Yukiyo, "US Army Veteran Gives Interview on 1959 Naha Accidental Nuke Missile Firing," *Ryukyu Shimpo*, October 26, 2017, http://english.ryukyushimpo .jp/2017/11/01/27963/ (accessed June 12, 2019). On Okinawa throughout the 1960s, other problems occurred during missile training drills. Author M. D. Morris noted how villagers complained about the Annual Service Practice, during which dozens of Hawk and Nike missiles were test fired from Bolo Point in western Okinawa. As well as the fishing restrictions, noise, and "brush fires caused by the falling of burning solid fuel," he notes villagers were angered by the $10 million the eight-day practice required, a "sum the Okinawan VIPs feel could much better be spent, even in part, on their health and welfare" (M. D. Morris, *Okinawa: A Tiger by the Tail* [New York City: Hawthorn, 1968], 97–98).

36. David E. Sanger, "A Missing H-Bomb Ruffles Japanese," *New York Times*, May 11, 1989, https://www.nytimes.com/1989/05/11/world/a-missing-h-bomb-ruffles -japanese.html (accessed June 12, 2019).

37. Office of the Historian, "Foreign Relations of the United States, 1964–1968, Volume XXIX, Part 2, Japan: 121. Editorial Note," *Office of the Historian*, https:// history.state.gov/historicaldocuments/frus1964-68v29p2/d121 (accessed June 12, 2019); "Alleged Radioactive Pollution from SSNs," electronic telegram, *US Embassy Tokyo*, February 27, 1974, https://wikileaks.org/plusd/cables/1974TOKYO02609_b .html (accessed June 12, 2019).

38. "Operational Report—Lessons Learned, Headquarters, 2d Logistical Command, Period Ending 31 October 1969."

39. "Nerve Gas Incident on Okinawa—Memorandum from Chief North Asia Branch to Director of Current Intelligence," Central Intelligence Agency, July 18, 1969.

40. Quoted in Havens, *Fire across the Sea*, 193. For a detailed exploration of the reversion agreement, see Gavan McCormack and Satoko Oka Norimatsu, *Resistant Islands: Okinawa Confronts Japan and the United States* (Lanham, Md.: Rowman and Littlefield, 2012), 55–62.

41. "History of the Custody and Deployment of Nuclear Weapons."

42. In November 1969, Japanese Foreign Ministry official Tanaka Hiroto informed the US national security adviser, Henry Kissinger, that Tokyo would "have no disagreement with the US" on bringing nuclear weapons back to the island. According to Prime Minister Sato Eisaku's special envoy, Wakaizumi Kei, in November 1969, the US and Japanese governments agreed such weapons could be deployed "in time of great emergency" to bases at Naha, Kadena, and Henoko ("Japan Officially Gave US Consent to Bring in Nuclear Weapons Ahead of Okinawa Reversion Accord: Document," *Japan Times*, August 14, 2017, https://www.japan times.co.jp/news/2017/08/14/national/history/japan-officially-gave-u-s-consent -bring-nukes-ahead-okinawa-reversion-accord-document/ [accessed June 12, 2019]; Steve Rabson, "Okinawa's Henoko Was a 'Storage Location' for Nuclear Weapons: Published Accounts," *Asia-Pacific Journal* 11, issue 1, no. 6 [January 14, 2013]: 1–5, https://apjjf.org/-Steve-Rabson/3884/article.pdf [accessed June 12, 2019]).

43. Thomas H. Barksdale and Marian W. Jones, "Rice Blast Epiphytology," US Army Biological Laboratories Fort Detrick, June 1965, 52.

44. Sheldon H. Harris, *Factories of Death: Japanese Biological Warfare, 1932–1945, and the American Cover-up* (London: Routledge, 1994), 233.

45. "Project 112 Fact Sheets," *Military Health System*, https://health.mil /Military-Health-Topics/Health-Readiness/Environmental-Exposures/Project -112-SHAD/Fact-Sheets (accessed June 14, 2019).

46. "Organizational History: 267th Chemical Company," US Army, March 26, 1966; "Overseas Storage of Chemical Agents/Munitions. Enclosure B: Discussion Pacific," Chief of Staff, US Army, July 22, 1969, CSAM 180-69.

47. Jon Mitchell, "Were US Marines Used as Guinea Pigs on Okinawa?" *Asia-Pacific Journal* 10, issue 51, no. 2 (December 17, 2012): 1–7, https://apjjf .org/2012/10/51/Jon-Mitchell/3868/article.html (accessed June 12, 2019).

48. Richard A. Hunt, "Melvin Laird and the Foundation of the Post-Vietnam Military 1969–1973," Historical Office, Office of the Secretary of Defense, Washington, D.C., 2015, 358–59. "Memorandum for the Assistant Secretary of Defense (System Analysis). Subject: Relocation of Chemical Munitions from Okinawa to Johnston Island," Joint Chiefs of Staff, January 9, 1971, DJSM 21-71.

49. "Trip Report," DTC 69-14, Tech Escort Support, March 27–29, 1969, 1 Lt. P. C. Spencer, 313 AD/DMW, April 30, 1969, in "History 400 MMS," January–June 1969, and Interview with Lt. Col P. C. Spencer, USAF, Ret., July 3, 1998, quoted in "Air Force History Report on Operation Tailwind," Air Force History Support Office, July 16, 1998, 15–16. In September 1970, US Special Forces allegedly used sarin agent during Operation Tailwind in Laos. When the allegations were reported by CNN and *Time* magazine in 1998, they were vehemently denied by the Pentagon, which asserted the chemicals used had been CS gas.

50. "Memorandum for the Assistant Secretary of Defense (System Analysis). Subject: Relocation of Chemical Munitions from Okinawa to Johnston Island," Joint Chiefs of Staff, January 9, 1971, DJSM 21-71.

51. James B. Lampert, US Military History Institute, Senior Officers Debriefing Papers, James and Williams Belote Papers; "Nerve Gas Accident: Okinawa Mishap Bears Overseas Deployment of Chemical Weapons," *Wall Street Journal*, July 18, 1969, 1.

52. "Nerve Gas Incident on Okinawa—Memorandum from Chief North Asia Branch to Director of Current Intelligence."

53. Jon Mitchell, "Ex-MP Revisits Okinawa's Koza Riot," *Japan Times*, January 8, 2011, https://www.japantimes.co.jp/community/2011/01/08/general/ex-mp-revisits -okinawas-koza-riot/ (accessed June 12, 2019).

54. Much of the following was initially revealed in Jon Mitchell, "Operation Red Hat: Chemical Weapons and the Pentagon Smokescreen on Okinawa," *Asia-Pacific Journal* 11, issue 30, no. 1 (August 5, 2013): 1–12, https://apjjf.org/2013/11/21/Jon -Mitchell/3975/article.html (accessed June 12, 2019).

55. Quoted in Mitchell, "Operation Red Hat," 2013.

56. Shimabukuro Ryota, "Forty Years after the Removal Operation of Stocks of US Poison Gas, a Representative of Residents Says, 'We Live Still Close to the Munitions Depot,'" *Ryukyu Shimpo*, July 15, 2011, http://english.ryukyushimpo .jp/2011/07/25/2018/ (accessed June 12, 2019); e-mail correspondence with Lindsay Peterson, October 2013.

57. "Shouwa 48 nen no 'Kyuugun dokugasu dan nado no zenkoku chousa' foro-appu chousa houkokusho," *Ministry of Environment*, November 2003, https://www .env.go.jp/chemi/report/h15-02/262 (accessed June 12, 2019).

58. "Analysis of Chemical Weapons Decontamination Waste from Old Ton Containers from Johnston Atoll Using Multiple Analytical Methods," *Environmental Science and Technology* 33, no. 13 (1999): 2,157–62.

59. Quoted in Mitchell, "Operation Red Hat," 2013.

## CHAPTER 4: MILITARY HERBICIDES, VIETNAM, AND OKINAWA

1. For a simple explanation of the history behind herbicide research, see David Zierler, *The Invention of Ecocide: Agent Orange, Vietnam, and the Scientists who Changed the Way we Think about the Environment*, (Athens: University of Georgia Press, 2011) 33–47.

2. Jonathan B. Tucker, *War of Nerves: Chemical Warfare from World War I to Al-Qaeda* (New York: Anchor Books, 2006), 102; William A. Buckingham, *Operation Ranch Hand: The Air Force and Herbicides in Southeast Asia, 1961–1971* (Washington, D.C.: Office of Air Force History, 1982), 4; Alastair Hay, *The Chemical Scythe: Lessons of 2,4,5-T and Dioxin* (New York: Springer US, 1982), 149;

William H. Cunliffe, *Select Documents on Japanese War Crimes and Japanese Biological Warfare, 1934–2006* (Washington, D.C.: National Archives and Records Administration, 2007), 47.

3. Letter from Philip W. Grone, principal assistant deputy undersecretary of defense (installations and environment), to Congressman Lane Evans, September 23, 2003.

4. Thomas H. Barksdale and Marian W. Jones, "Rice Blast Epiphytology," U.S. Army Biological Laboratories Fort Detrick, June 1965; Heianna Sumiyo, "Hokubu no karehazai sampu," *Okinawa Times*, September 6, 2011, 1. Some military officials also supported the usage of rice blast in Vietnam but the bioweapon was not deployed. (Barry Weisberg, *Ecocide in Indochina: The Ecology of War* [San Francisco, Calif.: Canfield Press, 1970], 62).

5. Quoted in Weisberg, *Ecocide in Indochina*, 6.

6. Buckingham, *Operation Ranch Hand*, 26–31.

7. Jeanne Stellman, et al. "The Extent and Patterns of Usage of Agent Orange and Other Herbicides in Vietnam," *Nature* 422, no. 681 (May 2003): 685.

8. Philip Jones Griffiths, *Agent Orange: "Collateral Damage" in Vietnam* (London: Trolley Limited, 2003), 18, 169.

9. "Herbicides, Pest Control Agents, and Disinfectants," Department of the Army Supply Bulletin, September 18, 1968.

10. Jeanne Stellman, et al. "The Extent and Patterns of Usage of Agent Orange and Other Herbicides in Vietnam," *Nature* 422, no. 681 (May 2003): 681–87; "Agent Orange: Actions Needed to Improve Accuracy and Communication of Information on Testing and Storage Locations" (Washington, D.C.: United States Government Accountability Office, November 2018), GAO-19-24, 26.

11. Quoted in Buckingham, *Operation Ranch Hand*, 33; quoted in Fred A. Wilcox, *Scorched Earth: Legacies of Chemical Warfare in Vietnam* (New York: Seven Stories Press, 2011), 21.

12. Weisberg, *Ecocide in Indochina*, 19.

13. Wilcox, *Scorched Earth*, 11; Weisberg, *Ecocide in Indochina*, 8.

14. Michael F. Martin, "Vietnamese Victims of Agent Orange and US–Vietnam Relations" (Washington, D.C.: Congressional Research Service, 2009), 22.

15. Tuyet, Le Thi Nham and Annika Johansson, "Impact of Chemical Warfare with Agent Orange on Women's Reproductive Lives in Vietnam: A Pilot Study," *Reproductive Health Matters* 9, No. 18, November 2001, 156–164.

16. Wilcox, *Scorched Earth*, 109; Weisberg, *Ecocide in Indochina*, 59.

17. Weisberg, *Ecocide in Indochina*, 60.

18. "Dioxins and Their Effects on Human Health," World Health Organization, October 4, 2016, https://www.who.int/news-room/fact-sheets/detail/dioxins-and -their-effects-on-human-health (accessed June 14, 2019).

19. Stellman et al., "The Extent and Patterns of Usage of Agent Orange and Other Herbicides in Vietnam," 681.

20. Wilcox, *Scorched Earth*, 124.

21. Peter Davis (director), *Hearts and Minds*, Rialto Pictures, 1974.

22. Zierler, *The Invention of Ecocide*, 122–23.

23. Fred A. Wilcox, *Waiting for an Army to Die: The Tragedy of Agent Orange* (Santa Ana, Calif.: Seven Locks Press, 1989), 82; Wilcox, *Scorched Earth*, 34.

24. Wilcox, *Scorched Earth*, 69.

25. Wilcox, *Scorched Earth*, 85–104; "Agent Orange," *Dow Chemical*, https://corporate.dow.com/en-us/about/issues-and-challenges/agent-orange (accessed June 14, 2019); "Agent Orange: Background on Monsanto's Involvement," *Monsanto*, https://monsanto.com/company/media/statements/agent-orange-background/ (accessed June 14, 2019).

26. One of the most comprehensive explorations of government malfeasance in relation to Agent Orange health studies is "Report to Secretary of the Department of Veterans Affairs on the Association between Adverse Health Effects and Exposure to Agent Orange," E. R. Zumwalt Jr, May 5, 1990.

27. Charles Ornstein and Mike Hixenbaugh, "Dr. Orange: The Secret Nemesis of Sick Vets," *ProPublica and Virginian-Pilot*, October 26, 2016, https://www.propublica.org/article/alvin-young-agent-orange-va-military-benefits (accessed June 14, 2019).

28. Ornstein and Hixenbaugh, "Dr. Orange: The Secret Nemesis of Sick Vets."

29. "Agent Orange: Actions Needed to Improve Accuracy and Communication of Information on Testing and Storage Locations," 16.

30. "Agent Orange: Actions Needed to Improve Accuracy and Communication of Information on Testing and Storage Locations," 14.

31. Martin, "Vietnamese Victims of Agent Orange and US–Vietnam Relations," 2.

32. Quoted in Wilcox, *Scorched Earth*, 91; quoted in Charles Waugh and Huy Lien, *Family of Fallen Leaves: Stories of Agent Orange by Vietnamese Writers* (Athens: University of Georgia Press, 2010), 2.

33. "Agent Orange Cleanup in Southern Vietnam," *NHK World*, aired April 21, 2019; "Assessment of Dioxin Contamination in the Environment and Human Population in the Vicinity of Da Nang Airbase, Vietnam," *Hatfield Consultants*, April 2007, https://www.hatfieldgroup.com/wp-content/uploads/AgentOrange Reports/DANDI1283/DANDI1283_Summary%20Document%20v4_protected .pdf (accessed June 14, 2019).

34. "Agent Orange Cleanup in Southern Vietnam," *NHK World*.

35. The VA's list is available at https://www.publichealth.va.gov/docs/agent orange/dod_herbicides_outside_vietnam.pdf (accessed June 14, 2019); "Agent Orange: Actions Needed to Improve Accuracy and Communication of Information on Testing and Storage Locations."

36. Jon Mitchell, "Evidence for Agent Orange on Okinawa," *Japan Times*, April 12, 2011, https://www.japantimes.co.jp/community/2011/04/12/issues/evidence-for -agent-orange-on-okinawa/ (accessed June 14, 2019).

37. Jon Mitchell, "Agent Orange Revelations Raise Futenma Stakes," *Japan Times*, October 18, 2011, https://www.japantimes.co.jp/community/2011/10/18 /issues/agent-orange-revelations-raise-futenma-stakes/ (accessed June 14, 2019).

38. "Karehazai o abita shima: Betonamu to okinawa, moto beigunjin no shougen," *Ryukyu Asahi Housou*, aired May 15, 2012; Jon Mitchell, "Okinawa Vet Blames Cancer on Defoliant," *Japan Times*, August 24, 2011, https://www.japantimes.co.jp/news/2011/08/24/national/okinawa-vet-blames-cancer-on-defoliant (accessed June 14, 2019).

39. Jon Mitchell, "US Vet Pries Lid Off Agent Orange Denials," *Japan Times*, April 15, 2012, https://www.japantimes.co.jp/news/2012/04/15/national/u-s-vet-pries-lid-off-agent-orange-denials/ (accessed June 14, 2019).

40. Jon Mitchell, "Agent Orange Buried on Okinawa, Vet Says," *Japan Times*, August 13, 2011, https://www.japantimes.co.jp/news/2011/08/13/national/agent-orange-buried-on-okinawa-vet-says (accessed June 14, 2019).

41. Darrow RA, ed. (1971): Historical, Logistical, Political and Technical Aspects of the Herbicide/Defoliant Program, 1967-1971. A Resume of the Activities of the Subcommittee on Defoliation/Anticrop Systems (Vegetation Control Subcommittee) for the Joint Technical Coordinating Group/Chemical-Biological. Plant Sciences Laboratories, US Army Chemical Corps, Fort Detrick, Frederick MD, September 1971, 48; "An Ecological Assessment of Johnston Atoll," United States Army Chemical Materials Agency, 2012, 4.

42. "Arsenic Poisoning of Beef Cattle—Memorandum for Public Health and Welfare Department," United States Civil Administration of the Ryukyu Islands, January 3, 1962.

43. "Karehazai o abita shima: 'Akuma no shima' to yobareta Okinawa," *TV Asahi*, aired October 9, 2012.

44. "Nerve Gas Incident on Okinawa—Memorandum from Chief North Asia Branch to Director of Current Intelligence," Central Intelligence Agency, July 18, 1969.

45. Jon Mitchell, "Beggars' Belief: The Farmers' Resistance Movement on Iejima Island, Okinawa," *Asia-Pacific Journal* 8, issue 23, no. 2 (June 7, 2010): 1–7, https://apjjf.org/-Jon-Mitchell/3370/article.pdf (accessed June 14, 2019).

46. "Beigun ga kareha sakusen," *Okinawa Times*, October 31, 1973, 1.

47. Jon Mitchell, "Agent Orange at Okinawa's Futenma Base in 1980s," *Asia-Pacific Journal* 10, issue 25, no. 3 (June 18, 2012): 1–6, https://apjjf.org/-Jon-Mitchell/3773/article.pdf (accessed June 14, 2019).

48. "Japan: Okinawa TV Serial Report Views Alleged Use of Agent Orange on US Bases," Central Intelligence Agency, January 15, 2012.

49. E-mail from Maj. Neal Fisher, USMC, Dep. Dir. Public Affairs HQ, US Forces Japan, October 6, 2011.

50. Letter from the Ministry of Foreign Affairs to the mayor of Nago City, September 18, 2012.

51. A. L. Young Consulting, Inc., "Investigations into Allegations of Herbicide Orange on Okinawa, Japan," 2013, 2.

52. "Daiokishin, kijunchi no 2.1 man bai, okinawa shi no doramu kan," *Ryukyu Shimpo*, June 30, 2016, https://ryukyushimpo.jp/news/prentry-245003.html (accessed June 14, 2019).

53. Travis J. Tritten, "Expert: Chemicals Found on Okinawa Likely Not Agent Orange," *Stars and Stripes*, August 15, 2013, https://www.stripes.com/news/pacific /expert-chemicals-found-on-okinawa-likely-not-agent-orange-1.235489 (accessed June 14, 2019); video of Kadena's commander comparing the barrels to empty tomato sauce cans was uploaded by the USAF to https://www.youtube.com /watch?v=Grf1LzXHpEk but has since been removed; "Fact Sheet: Dioxins and Dioxin-like Substances," Kadena Air Base, February 14, 2014.

54. Matthew M. Burke and Chiyomi Sumida, "Japanese Report: No Evidence of Agent Orange in Barrels on Okinawa," *Stars and Stripes*, July 24, 2014, https:// www.stripes.com/japanese-report-no-evidence-of-agent-orange-in-barrels-on-oki nawa-1.294966 (accessed June 14, 2019).

55. Wayne Dwernychuk, "Denials of Defoliant at Former US Base Site in Okinawa Fly in the Face of Science," *Japan Times*, August 26, 2013, https://www .japantimes.co.jp/community/2013/08/26/voices/denials-of-defoliant-at-former-u-s -base-site-in-okinawa-fly-in-the-face-of-science/ (accessed June 14, 2019).

56. "Himei o ageru tochi 1: Okinawa to betonamu osen no kyoutsuten," *Ryukyu Asahi Housou*, aired September 10, 2013, https://www.qab.co.jp /news/2013091046168.html (accessed June 14, 2019).

57. Jon Mitchell, "Kadena Moms Demand Truth," *Japan Times,* January 21, 2014, https://www.japantimes.co.jp/news/2014/01/21/national/kadena-moms -demand-truth (accessed June 14, 2019).

58. Jon Mitchell, "Contamination at Largest US Air Force Base in Asia: Kadena, Okinawa," *Asia-Pacific Journal*, 14, issue 9, no. 1 (May 1, 2016): 1–15, https://apjjf .org/2016/09/Mitchell.html (accessed June 14, 2019).

59. Letter from the Ministry of Foreign Affairs to the mayor of Nago City, September 18, 2012. The 1998 BVA ruling is available at https://www.va.gov/vetapp98 /files1/9800877.txt; the 2008 BVA decision is available at https://www.va.gov /vetapp08/files4/0831082.txt (both accessed June 14, 2019).

60. The 2013 decision is available at https://www.va.gov/vetapp13/Files4/1332861 .txt; the 2017 ruling can be read at https://www.va.gov/vetapp17/files6/1731591.txt (both accessed June 14, 2019).

61. "Agent Orange: Actions Needed to Improve Accuracy and Communication of Information on Testing and Storage Locations."

62. "Beigun dojou osen chousa, nihon futan 9.7 oku en," *Okinawa Times*, July 17, 2016, https://www.okinawatimes.co.jp/articles/-/54427 (accessed June 14, 2019).

## CHAPTER 5: POLLUTING WITH IMPUNITY

1. Neta C. Crawford, "Pentagon Fuel Use, Climate Change, and the Costs of War," *Boston University, Costs of War Project*, June 12, 2019, https://watson .brown.edu/costsofwar/files/cow/imce/papers/2019/Pentagon%20Fuel%20 Use,%20Climate%20Change%20and%20the%20Costs%20of%20War%20Final

.pdf (accessed July 1, 2019); "US Department of Defense Is the Worst Polluter on the Planet," *Project Censored*, October 2, 2010, https://www.projectcensored.org/2 -us-department-of-defense-is-the-worst-polluter-on-the-planet/ (accessed July 1, 2019); Emerson Urry, "The Department of Defense Is the Third Largest Polluter of US Waterways," *Truthout*, February 15, 2016, https://truthout.org/articles/the -department-of-defense-is-the-third-largest-polluter-of-us-waterways/ (accessed July 1, 2019); Joshua Frank, "The Pentagon Is Poisoning Your Drinking Water," *Counterpunch*, August 25, 2017, https://www.counterpunch.org/2017/08/25/the -pentagon-is-poisoning-your-drinking-water/ (accessed July 1, 2019); Dahr Jamail, "Naval Exercises Add Trillions of Pieces of Plastic Debris to Oceans," *Truthout*, March 15, 2017, https://truthout.org/articles/naval-exercises-add-trillions-of-pieces -of-plastic-debris-to-oceans/ (accessed July 1, 2019). Also, for an overview, see Barry Sanders, *The Green Zone: The Environmental Costs of Militarism* (Oakland, Calif.: AK Press, 2009).

2. Alexander Nazaryan, "The US Department of Defense Is One of the World's Biggest Polluters," *Newsweek*, July 17, 2014, https://www.newsweek .com/2014/07/25/us-department-defence-one-worlds-biggest-polluters-259456 .html (accessed July 1, 2019); Abrahm Lustgarten, "Open Burns, Ill Winds," *Pro-Publica*, July 20, 2017, https://features.propublica.org/military-pollution/military -pollution-open-burns-radford-virginia/ (accessed July 1, 2019).

3. Ken Miller, "Pentagon Says Environmental Mess Will Cost $25 Billion," *Gannett News Service*, May 13, 1993, 1; Thomas E. Baca, deputy assistant secretary of defense, Department of Defense Environmental Programs: Hearings before the Readiness Subcommittee, the Environmental Restoration Panel, and the Department of Energy Defense Nuclear Facilities Panel of the House Committee on Armed Services, 1991.

4. "Vets Exposed to Contaminated Water May Now Apply for Disability Benefits," *CBS News*, January 13, 2017, https://www.cbsnews.com/news/camp-lejeune -contaminated-water-veterans-benefits/ (accessed July 1, 2019).

5. Kelly Humphrey, "A Poisoned Legacy: Contractors Who Worked at Eglin's Agent Orange Spray Fields Still Live with Its Effects," *NWF Daily News*, August 22, 2015, https://www.nwfdailynews.com/article/20150822/NEWS/150829714 (accessed July 1, 2019); "Eglin Air Force Base: Potential Exposure Pathways," Agency of Toxic Substances and Disease Registry, October 21, 2009.

6. Seth Shulman, *The Threat at Home: Confronting the Toxic Legacy of the US Military* (Boston: Beacon, 1992), xi.

7. "Superfund Site Overview: Edwards Air Force Base," Environmental Protection Agency, https://cumulis.epa.gov/supercpad/cursites/csitinfo.cfm?id=0902725 (accessed February 12, 2018).

8. Tara Copp, "Why Women Were Told, 'Don't Get Pregnant at George Air Force Base,'" *Military Times*, June 19, 2018, https://www.militarytimes.com/news /your-military/2018/06/20/why-women-were-told-dont-get-pregnant-at-george-air -force-base/ (accessed July 1, 2019).

9. "Superfund Site Overview: Pearl Harbor Naval Complex," Environmental Protection Agency, https://cumulis.epa.gov/supercpad/cursites/csitinfo.cfm ?id=0904481 (accessed June 14, 2019); "Why Hawaii Wants the Navy to Move Its Fuel Tanks," *Military Times*, April 13, 2014, https://www.navytimes.com /news/your-navy/2019/04/14/why-hawaii-wants-the-navy-to-move-its-fuel-tanks/ (accessed July 1, 2019); Rob Perez, "Deal in Military Housing Lawsuit Kept Private," *Honolulu Star-Advertiser*, November 7, 2016, https://www.staradvertiser.com/2016/11/07/hawaii-news/mums-the-word-on-contamination-deal/ (accessed July 1, 2019).

10. Lustgarten, "Open Burns, Ill Winds."

11. John M. R. Bull, "The Deadliness Below," *Daily Press*, October 30, 2005, https://www.dailypress.com/news/dp-02761sy0oct30-story.html (accessed July 1, 2019). In August 2016, off the New Jersey coast, one fisherman was injured and hundreds of cases of clam chowder had to be destroyed following fears they had been exposed to mustard agent (Associated Press, "Clammer Is Burned Dredging Up Old Bomb Off N.J. Coast," *NJ.com*, August 12, 2016, https://www.nj.com/ocean/2016/08/clammer_is_injured_dredging_up_old_bomb_chowder_ge.html [accessed July 1, 2019]).

12. Shulman, *The Threat at Home*, 28–29.

13. In 1997, the EPA ordered the halt of live-fire training at Camp Edwards, Massachusetts. Previous use had contaminated the land there with unexploded munitions that leaked explosive components into the ground, including some linked to cancer. Also at Camp Edwards, the EPA ordered the military to begin cleanup of tens of thousands of kilograms of lead that had accumulated from firing small arms ("EPA Orders Further Training Restrictions and Cleanup at Camp Edwards," Environmental Protection Agency, April 10, 1997).

14. Shulman, *The Threat at Home*, 129–35; Robert F. Durant, *The Greening of the US Military: Environmental Policy, National Security, and Organizational Change* (Washington, D.C.: Georgetown University Press, 2007), Loc. 810.

15. John B. Wells, "We Have a Duty to Take Care of Our Veterans, but We're Letting Them Die," *The Hill*, November 16, 2017, https://thehill.com/opinion/national-security/360622-we-have-a-constitutional-duty-to-take-care-of-our-veterans-but-were (accessed July 1, 2019).

16. Durant, *The Greening of the US Military*, Loc. 145.

17. In 2004, the assistant attorney general for Colorado—home of the Rocky Mountain Arsenal's "most toxic mile on the planet"—lamented that the military thinks "they're beyond accountability" (Peter Eiser, "Pollution Cleanups Pit Pentagon against Regulators," *USA Today*, October 14, 2004, http://www.usatoday.com/news/nation/2004-10-14-cover-pollution_x.htm [accessed July 1, 2019]).

18. Durant, *The Greening of the US Military*, Loc. 1875.

19. Durant, *The Greening of the US Military*, Loc. 3859.

20. Shulman, *The Threat at Home*, 118; Durant, *The Greening of the US Military*, Loc. 3714; Jeffrey St. Clair and Joshua Frank, "The Pentagon's Toxic Leg-

acy," *Counterpunch*, May 12, 2008, https://www.counterpunch.org/2008/05/12/the
-pentagon-s-toxic-legacy/ (accessed July 1, 2019).

21. "Military Base Realignments and Closures: DoD Has Improved Environ-
mental Cleanup Reporting but Should Obtain and Share More Information,"
*Government Accountability Office*, January 19, 2017, https://www.gao.gov/products
/GAO-17-151 (accessed July 1, 2019).

22. Teresa Albor, "US Leaves Toxins at Subic Navy Base," *Christian Science
Monitor*, November 24, 1992, https://www.csmonitor.com/1992/1124/24012.html
(accessed July 1, 2019).

23. US Department of Defense, Office of the Inspector General, *Final Report:
Review of Hazardous Material/Hazardous Waste Management within the Department of Defense*, October 15, 1985–February 21, 1986 (Washington, D.C.: US
Department of Defense, Office of the Inspector General, 1986), quoted in Seth
Shulman, *The Threat at Home*, 108. In response to a FOIA request for the report
to the US Department of Defense, Office of the Inspector General, I was told the
office had no records in March 2017.

24. US General Accounting Office, "Hazardous Waste: Management Problems
at DoD's Overseas Installations," September 1986. In April 2017, I initiated a man-
datory declassification review for the report, but as of January 2020, it had not yet
been processed.

25. US General Accounting Office, "Hazardous Waste: Management Problems
Continue at Overseas Military Bases," August 1991, 32, 4.

26. GAO, "Hazardous Waste," 1991.

27. By far the most comprehensive—and readable—exploration of this postwar
period is John W. Dower, *Embracing Defeat: Japan in the Wake of World War II*
(New York: W.W. Norton & Company, 1999).

28. For an excellent visual overview of this period, see "Tokyo 1960: Days of
Rage and Grief," *MIT Visualizing Cultures*, https://visualizingcultures.mit.edu
/tokyo_1960/anp2_essay01.html (accessed July 1, 2019).

29. For discussion of CIA support for the LDP, see Tim Weiner, "CIA Spent
Millions to Support Japanese Right in 50s and 60s," *New York Times*, October
9, 1994, https://www.nytimes.com/1994/10/09/world/cia-spent-millions-to-support-
japanese-right-in-50-s-and-60-s.html (accessed June 14, 2019) and Tim Weiner,
*Legacy of Ashes: The History of the CIA* (New York: Doubleday, 2007), 133–140.

30. "Agreement under Article VI of The Treaty of Mutual Cooperation and Security between Japan and the United States of America, Regarding Facilities and
Areas and the Status of United States Armed Forces in Japan," https://www.mofa
.go.jp/region/n-america/us/q&a/ref/2.html (accessed July 1, 2019).

31. International Civil Society Workshop Secretariat (ed.), "Civil Society Workshop Report" (Ginowan: Okinawa International University, 2009), 15.

32. In Japan, aircraft at Kadena, Atsugi, and Iwakuni air bases have caused such
problems as in-flight fuel leaks, emissions of carcinogenic gases, and noise pollution. Spills from military ships, for example, radioactive contamination, have been

a constant cause of concern on both Okinawa and mainland Japan. In 2008, the US government informed Tokyo that its nuclear submarine, USS *Houston*, had been leaking radiation-contaminated cooling water. Between June 2006 and April 2008, the vessel had visited White Beach five times, Sasebo five times, and Yokosuka once. The military claimed the amounts of radiation spilled were too low to damage human health ("US Nuclear-Powered Submarine Has Leaked Radioactive Water for More Than Two Years," *Japan Press Weekly*, August 8, 2008, https://www.japan-press.co.jp/2008/2585/radioactive_leak_4.html [accessed June 14, 2019]). Furthermore, in January 2015, the US Navy's amphibious assault ship USS *Bonhomme Richard* dumped 151,416 liters of waste water into Okinawa's Nakagusuku Bay. The discharge, originating from onboard lavatories, medical facilities, and laundries, was not reported to Japanese authorities for three days, by which time it was too late to remediate any damage. Details of the incident only came to light via a FOIA request ("Beigunkan umi ni osui 15 man rittoru suteru," *Okinawa Times*, January 17, 2017, https://www.okinawatimes.co.jp/articles/-/80119 [accessed June 14, 2019]).

33. "Joint Statement of Environmental Principles," *Japan Ministry of Foreign Affairs*, September 11, 2000, https://www.mofa.go.jp/mofaj/area/usa/sfa/rem_env_01en.html (accessed June 14, 2019).

34. "Joint Announcement on a Framework Regarding Environmental Stewardship at US Armed Forces Facilities and Areas in Japan," *US Embassy, Tokyo*, December 25, 2013, https://jp.usembassy.gov/joint-announcement-framework-regarding-environmental-stewardship-u-s-armed-forces-facilities-areas-japan/ (accessed June 14, 2019).

35. "Cooperation Concerning Environmental Matters," *Japan Ministry of Foreign Affairs*, September 28, 2015, https://www.mofa.go.jp/mofaj/files/000101626.pdf (accessed June 14, 2019).

36. Sharon K. Weiner, "DACS Working Paper: Environmental Concerns at US Overseas Military Installations," MIT Defense and Arms Control Studies Program, July 1992; Michael S. Darnell, "Ansbach Drinking Water Safe, but Chemical Cleanup Continues," *Stars and Stripes*, June 24, 2016, https://www.stripes.com/ansbach-drinking-water-safe-but-chemical-cleanup-continues-1.416131 (accessed June 14, 2019).

37. "[Editorial] Let's Not Get Left Cleaning Up US Military Bases' Polluted Mess," *Hankyoreh*, March 14, 2015, http://www.hani.co.kr/arti/english_edition/e_editorial/682232.html (accessed June 14, 2019); David Vine, *Base Nation: How US Military Bases Abroad Harm America and the World* (New York: Metropolitan Books, 2015), 143.

38. The most comprehensive source for information on the Shinkampo contamination incident at the NAFA incinerator is the website maintained by Daniel Larsen at http://nafatsugiincineratorgroup.weebly.com/.

39. US Congress, Senate, Hearing before the Committee on Veterans' Affairs, "VA/DoD Response to Certain Military Exposures," United States Senate, 111th Congress, First Session (Senate Hearing 111-437), October 8, 2009.

40. US Congress, Senate, Hearing before the Committee on Veterans' Affairs, "VA/DoD Response to Certain Military Exposures"; "Atsugi Base Dioxin Uproar: US Irritated at Slow Response," *Tokyo Shimbun*, February 6, 2000, 6.

41. "Captain's Call," *Skywriter*, October 10, 1997, 2.

42. Doug Struck, "Blowing Smoke in Japan: Incinerator Near Base Fouls Air and Relations with US," *Washington Post*, February 22, 2000, https://www.washingtonpost.com/wp-srv/WPcap/2000-02/22/035r-022200-idx.html (accessed June 14, 2019).

43. "Waste Incinerator in Atsugi, Japan," *US Department of Veterans Affairs*, https://www.publichealth.va.gov/exposures/sand-dust-particulates/atsugi.asp (accessed June 14, 2019).

44. Quoted in US Congress, Senate, Hearing before the Committee on Veterans' Affairs, "VA/DoD Response to Certain Military Exposures."

45. The GAO also criticized the navy's failure to notify those who may have been sickened. The Department of Defense contested the conclusion, arguing that current laws do "not require responsible parties to identify individuals who may have been exposed to contamination in the past" (US Government Accountability Office, *Defense Infrastructure: DoD Can Improve Its Response to Environmental Exposures on Military Installations*, GAO-12-412 [Washington, D.C.: May 2012], 49).

46. In 2005, the Koshiba POL Depot, Kanagawa Prefecture, closed after fifty-seven years of use as a US military fuel storage depot. Yokohama City planned to redevelop the area into a park, but the project was delayed by the subsequent discovery of contamination from benzene, lead, and arsenic. On land at the former Camp Asaka, Saitama Prefecture, lead contamination thirty times the safe level was announced in 2007. Also at that time, the municipality was planning to construct a park in the area ("Kyuu koshiba choyu shisetsu no dojou chousa, osen keni wa 3.6%," *Kanagawa Shimbun*, June 29, 2010, https://www.kanaloco.jp/article/entry-135546.html [accessed June 14, 2019]; "Asaka kichi atochi kijunchi 20 bai no namari kenshutsu," *Tokyo Shimbun*, May 16, 2007). Military lead pollution again caused public concerns in 2011, when it was revealed that land that had once been part of Tachikawa Air Base, Tokyo, contained the heavy metal at twenty-eight times the safe level. On the site, the city planned to build a facility to make school lunches; the discovery caused public alarm ("Kyuushoku chouriba yoteichi ni namari/ Tokyo tachikawa," *Shimbun Akahata*, January 21, 2011).

47. Shimabukuro Natsuko, "Kyuu beigun youchi no genjou kaifuku ni sosogareta nihon no 129 oku en," *Webronza*, November 18, 2018, https://webronza.asahi.com/politics/articles/2018111400003.html (accessed June 14, 2019).

48. US Government Accountability Office, *Overseas Presence: Issues Involved in Reducing the Impact of the US Military Presence on Okinawa*, GAO/NSIAD-98-66 (Washington, D.C.: March 1998), 7.

49. "Okinawa no haikibutsu fukushima ni/Beigun PCB shobun e," *Shimbun Akahata*, September 23, 2013, http://www.jcp.or.jp/akahata/aik13/2013-09-23/2013092301_04_1.html (accessed June 14, 2019); "Jieitai onna buntonkichi PCB osen busshitsu kengai e hanshutsu," *Ryukyu Asahi Housou*, aired November

12, 2013, https://www.qab.co.jp/news/2013111247738.html (accessed June 14, 2019). Not including the initial surveys and cleanup work, the cost of the shipment and disposal of the 1,794 barrels of waste came to 395 million yen ($3.6 million), of which the United States did not pay a single yen.

50.  Irvin Molotsky, "Admiral Has to Quit over His Comments on Okinawa Rape," *New York Times*, November 18, 1995, https://www.nytimes.com/1995/11/18/world /admiral-has-to-quit-over-his-comments-on-okinawa-rape.html (accessed June 14, 2019).

51.  Ikeda Yukihiko, Kyuma Fumio, William Perry, and Walter Mondale, "The Special Action Committee on Okinawa Final Report," *Japanese Ministry of Foreign Affairs*, December 2, 1996, https://www.mofa.go.jp/region/n-america/us/security /96saco1.html (accessed June 14, 2019).

52.  "Atochi no keizai kouka 28 bai/Kichi, hatten no ashikase ni," *Ryukyu Shimpo*, May 22, 2016, https://ryukyushimpo.jp/news/entry-283959.html (accessed June 14, 2019). Elsewhere, at the site of the CIA's former communications station at Yomitan, which was returned in 2006, the Okinawa Defense Bureau discovered high levels of lead and suspected oil contamination between November 2006 and February 2007 ("Senaha kijunchi koeru namari," *Ryukyu Asahi Housou*, aired February 26, 2007, https://www.qab.co.jp/news/200702265204.html [accessed June 14, 2019]).

53.  "Nishi futenma no dojou osen," *Ryukyu Shimpo*, May 24, 2016, https:// ryukyushimpo.jp/news/entry-285026.html (accessed June 14, 2019).

54.  "A Master Narratives Approach to Understanding Base Politics in Okinawa," Central Intelligence Agency, January 5, 2012. For commentary, see Jon Mitchell, "CIA: How to Shape Okinawan Public Opinion on the US Military Presence," *Asia-Pacific Journal* 16, issue 13, no. 5 (July 1, 2018): 1–10, https://apjjf.org/2018/13 /Mitchell.html (accessed June 14, 2019).

55.  "Japanese Gov't Prevented from Environmental Inspections on US Bases since FY2014," *Mainichi*, November 16, 2017, https://mainichi.jp/english/articles /20171116/p2a/00m/0na/008000c (accessed June 14, 2019).

## CHAPTER 6: OKINAWA: PARADISE LOST

1.  Travis Tritten, "State Dept. Official in Japan Fired Over Alleged Derogatory Remarks," *Stars and Stripes*, March 9, 2011, https://www.stripes.com/news/pacific /japan/state-dept-official-in-japan-fired-over-alleged-derogatory-remarks-1.137181 (accessed June 14, 2019); Jon Mitchell, "US Marines Official Dismissed Over Okinawan Protest Video Leak," *Japan Times*, March 23, 2015, https://www.japan times.co.jp/community/2015/03/23/issues/u-s-marines-official-dismissed-okinawa -protest-video-leak/ (accessed June 14, 2019); Zaha Yukiyo, "Commandant Neller Ignores History of Land Taken by US Military, Saying, 'There Were No People Living' Near Futenma When It Was Built," *Ryukyu Shimpo*, March 4, 2018, http:// english.ryukyushimpo.jp/2018/05/12/28809/ (accessed June 14, 2019).

2. "Department of Defense Announces Winners of the 2015 Environmental Awards," *Department of Defense*, April 21, 2015, https://dod.defense.gov/News/News-Releases/News-Release-View/Article/605471/department-of-defense-announces-winners-of-the-2015-environmental-awards/ (accessed June 12, 2019).

3. "Beigun kichi kankyou karute: Makiminato hokyuuchiku," *Okinawa Prefecture*, March 2017, https://www.pref.okinawa.jp/site/kankyo/seisaku/documents/54.pdf, 54-13 (accessed June 12, 2019). On February 12, 1976, another base worker fell seriously ill during pest extermination work. He was subsequently diagnosed as suffering from methyl bromide poisoning. Again in 2009, six Japanese workers fell ill following exposure to an unknown substance at a warehouse on the base, but in response to my FOIA request for reports on the accident, the USMC said they could not find any records (David Allen, "Hazmat Team Investigates Odor at Kinser Warehouse," *Stars and Stripes*, April 30, 2009, https://www.stripes.com/news/hazmat-team-investigates-odor-at-kinser-warehouse-1.90911 [accessed June 12, 2019]).

4. Jon Mitchell, "Pentagon Blocks Report on 'Toxic Contamination' at Base Outside Okinawa Capital," *Japan Times*, September 16, 2015, https://www.japantimes.co.jp/community/2015/09/16/issues/pentagon-blocks-report-toxic-contamination-base-outside-okinawa-capital/#.Xf01LXt7nIU (accessed June 12, 2019); "USFJ Talking Paper on Possible Toxic Contamination at Camp Kinser, Okinawa," United States Forces in Japan, July 30, 1993.

5. "Mongooses Near US Bases Have High PCB Levels," *Kyodo News*, August 19, 2013, https://www.japantimes.co.jp/news/2013/08/19/national/mongooses-near-u-s-bases-have-high-pcb-levels/#.Xf4lzXt7nIU (accessed June 12, 2019); "Habu kara kounoudo PCB to DDT," *Okinawa Times*, September 4, 2015, https://www.okinawatimes.co.jp/articles/-/18359 (accessed June 12, 2019); "Camp kinser fukin ni okeru teishitsu chousa no kekka houkoku ni tsuite," Okinawa Prefectural Government, October 20, 2016; "Camp kinser shuuhen no teishitsu chousa/Kuni no kankyou kijunchi shitamawaru," *Okinawa Times*, January 21, 2017, https://www.okinawatimes.co.jp/articles/-/80715 (accessed June 12, 2019).

6. Jon Mitchell, "Environmental Contamination at USMC Bases on Okinawa," *Asia-Pacific Journal* 15, issue 4, no. 2 (February 15, 2017): 1–7, https://apjjf.org/2017/04/Mitchell.html (accessed June 12, 2019). Inspections by Marine Corps Installations Command uncovered 110 violations in 2013, and, in 2017, another 106 violations; meanwhile, interim checks conducted by the Okinawa-stationed USMC in 2014, also found 188 violations. Breaches included mismanagement of hazardous materials like PCBs, asbestos, and pesticides, and violations involving wastewater, air emissions, and garbage disposal. Japanese authorities had not been informed about any of the cases (Jon Mitchell, "Zainichi kaiheitai kichi de kankyou ihan aitsuzuku," *Okinawa Times*, September 13, 2018, https://www.okinawatimes.co.jp/articles/-/314128 [accessed June 12, 2019]).

7. The actual number of accidents is certainly higher than 270, because large tracts of information had been redacted from the FOIA-released documents, and for some years no accident reports were released.

8. Negligence is a common cause of many of the accidents occurring on Okinawa's USMC bases. In September 2005, contractors at Camp Schwab accidentally cut a fuel line. The spill went unnoticed for four days and contaminated 120 meters of a river which flows into Oura Bay, home to endangered corals, seagrasses, and the near-extinct dugong. At Camp Hansen in November 2008, a marine washed "unknown POLs" into drains, which then flowed off the base near an elementary school; the military did not bother to inform the school of its pupils' potential exposure. In February, a marine living on Camp Foster poured cooking oil into a sink, blocking pipes and causing a nineteen-thousand-liter leak of sewage that flowed through local communities. According to one internal e-mail, the USAF environmental officer in charge of the cleanup did not reprimand the marine because he was afraid he might be beaten up.

9. "Beigun kichi kankyou karute: Camp Hansen," *Okinawa Prefecture*, March 2017, http://www.pref.okinawa.jp/site/kankyo/seisaku/documents/09.pdf, 9–14 (accessed June 12, 2019); "Beigun kichi kankyou karute: Hokubukunrenjou," *Okinawa Prefecture*, March 2017, https://www.pref.okinawa.jp/site/kankyo/seisaku /documents/01.pdf, 1–13 (accessed June 12, 2019).

10. "Integrated Natural Resources and Cultural Resources Management Plan: Marine Corps Base Camp Smedley D. Butler MCIPAC Installations, Okinawa, Japan," United States Marine Corps, April 2014.

11. In June 1997, for example, about five hundred thousand square meters burned following the ignition of gunpowder being removed from UXO; in December of that same year, a tracer round sparked a fire that burnt roughly 562,500 square meters ("Okinawa no beigun oyobi jieitai kichi," *Naha: Okinawa Prefectural Government*, 2016, https://www.pref.okinawa.jp/site/chijiko/kichitai /documents/4-2.pdf, 105 [accessed June 12, 2019]). Peace Depot data included in "UNEP International Civil Society Workshop on Environmental Norms and Military Activities," *Okinawa International University*, November 2009, http://un epcsows.oki-kan.net/Workshop_Report_files/Workshop%20Report%20Final%20 Opening.pdf, 14–15 (accessed June 12, 2019).

12. For the background on this issue, see (in English) David Allen and Chiyomi Sumida, "Skeet Range, Closed in 1999, Still a Hot Issue on Okinawa," *Stars and Stripes*, July 16, 2003, https://www.stripes.com/news/skeet-range-closed-in-1999 -still-a-hot-issue-on-okinawa-1.7613#.WXR-ZTOB3sk (accessed June 12, 2019). An in-depth exploration is available in Japanese in "Okinawa no beigun kichi," *Okinawa Prefecture*, March 2013, chapters 3 and 69, http://www.pref.okinawa.lg.jp /site/chijiko/kichitai/documents/dai3syou.pdf (accessed June 12, 2019).

13. Sakurai Kunitoshi, "Yogosaretamama henkansareru beigun kichi," *Webronza*, February 14, 2018, https://webronza.asahi.com/science/articles/2018021000001 .html (accessed June 12, 2019); Jon Mitchell, "Houshasei busshitsu youkakan zanryu," *Okinawa Times*, October 11, 2018, https://www.okinawatimes.co.jp/articles /-/328541 (accessed June 12, 2019).

14. Jon Mitchell, "How the US Military Spies on Okinawans and Me," *Japan Times*, October 19, 2016, https://www.japantimes.co.jp/community/2016/10/19

/issues/u-s-military-spies-okinawans/ (accessed June 12, 2019); "US Military in Okinawa Spy on Journalists," *Reporters Without Borders*, October 23, 2016, https://rsf.org/en/news/us-military-okinawa-spy-journalists (accessed June 12, 2019).

15. "Okinawa Cultural Awareness Training," United States Marine Corps, 2011, Slide 26. Also see: Jon Mitchell, "Okinawa: US Marines Corps Training Lectures Denigrate Local Residents, Hide Military Crimes," *Asia-Pacific Journal* 14, issue 13, no. 4 (July 1, 2016): 1–5, https://apjjf.org/2016/13/Mitchell.html (accessed June 12, 2019).

16. "Okinawa Cultural Awareness Training," United States Marine Corps, 2014, 14.

17. Amy Hagopian, Riyadh Lafta, Jenan Hassan, Scott Davis, Dana Mirick, and Tim Takaro, "Trends in Childhood Leukemia in Basrah, Iraq, 1993–2007," *American Journal of Public Health* 100, no. 6 (June 1, 2010): 1,081–87; Dahr Jamail, "Iraq: War's Legacy of Cancer," *Al Jazeera*, March 16, 2013, https://www.aljazeera.com/indepth/features/2013/03/2013315171951838638.html (accessed June 12, 2019).

18. "Technical Review of Tori Shima Surveys," United States Air Force, July 9, 1999.

19. "Integrated Natural Resources Management Plan FY 2010–2015 for Kadena Air Base, Okinawa, Japan," Center for Environmental Management of Military Lands, Colorado State University, September 2010, 175.

20. "Okinawa beikuugun kadena kichi/Rekka uran dan 40 manpatsu o hokan," *Mainichi Shinbun*, August 2, 2006, https://www.mainichi-msn.co.jp/kokusai/america/usa_c/news/20060802k0000m040167000c.html (Expired link); "Kinoko gumo no you na funen," *RBC Ryukyu Housou the News*, aired May 22, 2014.

21. For an overview of the FOIA-released documents, see Jon Mitchell, "Contamination at Largest US Air Force Base in Asia: Kadena, Okinawa," *Asia-Pacific Journal*, 14, issue 9, no. 1 (May 1, 2016): 1–15, https://apjjf.org/2016/09/Mitchell.html (accessed June 12, 2019). The documents also catalog frequent discoveries of UXO on or near Kadena Air Base. In January 2015, construction crews at a Patriot missile site accidentally dug up UXO containing almost two kilograms of white phosphorus. Fearing attempts to collect the contaminated earth might result in an explosion, the environmental officers left it in place, smoldering for more than a day. In June 2016, another smoking white phosphorus round was discovered in the ammunition depot, prompting evacuation before the ordnance could be cleared.

22. The 1993 contamination from the incinerator measured 14,000 mg/kg; the 1994 contamination exceeded 500 mg/kg. The Japanese government's cleanup standard for lead contamination in soil is 150 mg/kg. Japan has no standard for agricultural land, but in Germany, for example, the maximum level permitted is 100 mg/kg. ("Consultative Letter (CL), AL/OE-CL-1994-0183, Sampling Recommendations to Determine the Extent of Lead Contamination in Soil Surrounding the Munitions Deactivation Furnace, Bldg 46808, Kadena AB, Okinawa, Japan," United States Air Force, January 16, 1995.)

23. "Consultative Letter (CL), AL/OE-1997-0006 Sand Blasting Media as a Hazardous Waste Stream," United States Air Force, February 20, 1997.

24. "Consultative Letter, IERA-DO-BR-CL-2000-0008, Asbestos Survey of Building 703, Kadena AB, Japan," United States Air Force, March 17, 2000.

25. "Twenty-Eight Japanese Confirmed with Asbestos Injuries from Working at US bases," *Kyodo News*, January 8, 2014, https://www.japantimes.co.jp /news/2014/01/08/national/28-japanese-confirmed-with-asbestos-injuries-from -working-at-u-s-bases/#.Xf47ZHt7nIU (accessed June 12, 2019).

26. "Military Housing Inspections – Japan" (Report No. DODIG-2014-121), Inspector General, US Department of Defense, September 30, 2014.

27. "Department of Defense Dependent Schools Lead Assessment Project," Department of the Air Force Pacific Air Forces, August 27, 2015.

28. At the time the pollution was discovered, Japan's cleanup standard for PCBs in soil stood at 3 ppm, whereas in the United States, it was 25 ppm. Today, Japan's regulations are much stricter, as low as 0.03 ppm, and the United States allows 25 ppm only for industrial areas in which people spend short amounts of time, lessening their risk of exposure. For more, see Jon Mitchell, "Military Contamination on Okinawa: PCBs and Agent Orange at Kadena Air Base," *Asia-Pacific Journal* 12, issue 12, no. 1 (March 24, 2014): 1–7, https://apjjf.org/2014/12/12/Jon-Mitchell/4097 /article.html (accessed June 12, 2019).

29. Suhas Chakma and Marianne Jensen (eds.), *Racism against Indigenous Peoples* (Copenhagen: International Work Group for Indigenous Affairs, 2001), 110; "Toxic Bases in the Pacific," *Nautilus Reports*, November 25, 2005, https://nautilus .org/apsnet/toxic-bases-in-the-pacific/ (accessed June 12, 2019).

30. "Beigun kichi kankyou karute: Kadena hikoujou," *Okinawa Prefecture*, March 2017, http://www.pref.okinawa.jp/site/kankyo/seisaku/documents/35.pdf (accessed June 12, 2019).

31. In October 2005, an airplane dumped fuel over busy Highway 58 after a valve malfunctioned; in May 2006, a similar failure caused the dumping of approximately two thousand liters of fuel over land. In July 2016, air sampling near Kadena Air Base also revealed levels of benzene and 1,3-butadiene—both carcinogens—higher than safe levels ("Kadena no taiki osen, kijunkoe," *Okinawa Times*, September 30, 2016, https://www.okinawatimes.co.jp/articles/-/64380 [accessed June 12, 2019]).

32. "Burden of Disease from Environmental Noise," World Health Organization: Regional Office for Europe, 2011.

33. Ito Kazuyuki, "Study: Jet Noise at US Kadena Base Triggers Ten Deaths a Year," *Asahi Shimbun*, March 14, 2019, http://www.asahi.com/ajw/articles /AJ201903140041.html (accessed June 12, 2019).

34. In 2019, a high court reduced the sum to 26.1 billion yen ($237.3 million) ("Compensation Order Over Okinawa Base Noise Upheld, But Amount Cut," *Japan Today*, September 11, 2019, https://japantoday.com/category/national /compensation-order-over-okinawa-base-noise-upheld-but-amount-cut-1 [accessed June 12, 2019]).

35. "Consultative Letter, AL-CL-1992-0110, Evaluation of Water Sampling Site 4, Kadena AB, JA," United States Air Force, August 14, 1992.

36. Another report from 1992, revealed that an industrial wastewater treatment system wasn't working, so it passed contaminants into the sewer system. These substances included methylene chloride, phenol, lead, chromium, and zinc—all of which exceeded allowable standards. ("Consultative Letter, AL-CL-1992-0118, Evaluation of Industrial Water Treatment Plant, Bldg 3448, Kadena AB, JA," United States Air Force, September 9, 1992.)

37. For example, in August 2011, 760 liters of diesel spilled into the Hija River when an operator abandoned a generator tank prior to the arrival of a typhoon. In June 2012, it took an engineer an hour and twenty minutes to respond to a large fuel spill because he was at the food court and could not hear his telephone ring. In February 2015, environmental teams failed to even respond to two accidents despite being alerted by emergency crews.

38. Mitchell, "Contamination at Largest US Air Force Base in Asia: Kadena, Okinawa."

39. Tom Roeder and Jakob Rodgers, "Toxic Legacy: Air Force Studies Dating Back Decades Show Danger of Foam That Contaminated Colorado Springs-Area Water," *Gazette*, October 23, 2016, https://gazette.com/health/toxic-legacy-air -force-studies-dating-back-decades-show-danger/article_024f688b-9f1e-5395 -9819-dc97cf71bf9d.html (accessed June 12, 2019); Melanie Benesh and Audrey Lothspeich, "Mapping PFAS Chemical Contamination at 206 US Military Sites," *Environmental Working Group*, July 19, 2019, https://www.ewg.org/research/pfas -chemicals-contaminate-us-military-sites (accessed January 6, 2020);"DuPont, 3M Concealed Evidence of PFAS Risks," *Union of Concerned Scientists*, https://www.uc susa.org/our-work/center-science-and-democracy/disinformation-playbook/dupont -3m-concealed-evidence-pfas (accessed June 14, 2019); Melanie Benesh and Audrey Lothspeich, "Mapping PFAS Chemical Contamination at 206 US Military Sites," *Environmental Working Group*, July 19, 2019, https://www.ewg.org/ research/pfas-chemicals-contaminate-us-military-sites (accessed January 6, 2020).

40. Maureen Sullivan, "Addressing Perfluorooctane Sulfonate (PFOS) and Perfluorooctanoic Acid (PFOA)," Deputy Assistant Secretary of Defense (Environment, Safety, and Occupational Health), March 2018. For data as of October 2019, see also Meghann Myers, "These Ninety Army Posts Have Contaminated Drinking Water," *Military Times*, September 11, 2019, https://www.militarytimes.com/news /your-military/2019/09/11/these-90-army-posts-have-contaminated-drinking -water/ (accessed June 14, 2019).

41. Gerald B. Silverman, "Glass Half-Full on State Solutions to Chemicals in Water," *Bloomberg Environment*, September 18, 2018, https://news.bloomberg environment.com/environment-and-energy/glass-half-full-on-state-solutions-to -chemicals-in-water-corrected (accessed June 14, 2019); Annie Snider, "White House, EPA Headed Off Chemical Pollution Study," *Politico*, May 14, 2018, https://www.politico.com/story/2018/05/14/emails-white-house-interfered-with-sci ence-study-536950 (accessed June 14, 2019).

42. Naval Criminal Investigation Service records reveal the release caused $92,381 in damages in the hangar, but the perpetrator was only punished with ninety days' confinement and a pay cut.

43. According to an accident report from August 1997, 3,028 liters of foam and water from a fire truck were poured into the base's drains in an incident described by environmental officers as "harmful to the water." More recently, in May 2015, a foam spill was described as "PFOS contaminated"—but even though it leaked off the base, the Japanese government was not informed.

44. "Beigun kichi shuuhen karasaikounoudo no yuugaibusshitsu o kenshutsu," *Okinawa Times*, April 24, 2019, https://www.okinawatimes.co.jp/articles/-/412770 (accessed June 14, 2019); "Kecchuu ni yuugaibusshitsu PFHxS," *Okinawa Times*, May 18, 2019, https://www.okinawatimes.co.jp/articles/-/421237 (accessed June 14, 2019); "PFOS/Taimo wa anzen/Okinawa ginowan de kyoudai chousa," *Ryukyu Shimpo*, May 27, 2019, https://ryukyushimpo.jp/news/entry-925358.html (accessed June 14, 2019).

45. "PFOS osen 'shitteita' ga kouhyou sezu," *Ryukyu Asahi Housou*, aired June 19, 2019, https://www.qab.co.jp/news/20190619116290.html (accessed June 30, 2019).

46. "Kadenakichi no PFOS osen, bei kijun no saidai 1 oku bai," *Okinawa Times*, January 11, 2019, https://www.okinawatimes.co.jp/articles/-/369927 (accessed June 19, 2019); "Japan/DET 3 KADENA AFB, JAPAN / EC18-040," Maxxam Analytics for United States Air Force, November 10, 2017; "Futenma hikoujou ni yuugai busshitsu, kounoudo de osen," *Okinawa Times*, October 27, 2018, https://www.okinawatimes.co.jp/articles/-/336091 (accessed June 19, 2019); "Certificate of Test Results, No.2015-02203-B01~04" Okinawa Prefecture Environment Science Center for United States Marine Corps, February 18, 2016.

47. "Awa shoukazai 142 ton 'kurashiki' ni," *Okinawa Times*, June 26, 2019, https://www.okinawatimes.co.jp/articles/-/437789, (accessed January 10, 2020).

## CHAPTER 7: JAPAN: CONTAMINATION, NUCLEAR DEALS, AND THE FUKUSHIMA MELTDOWNS

1. During World War II, NAFA had almost nine thousand 60-kilogram mustard agent bombs, and Yokosuka Naval Base had thirty thousand Red Agent munitions. In March 1955, at a US base in Chitose, Hokkaido, thirty workers and members of the Japanese Self-Defense Forces were injured during efforts to dispose of a tank of blister agent ("Shouwa 48 nen no 'Kyuugun dokugasu dan nado no zenkoku chousa' foroappu chousa houkokusho," *Ministry of the Environment*, November 2003, https://www.env.go.jp/chemi/report/h15-02/ [accessed June 14, 2019]).

2. "Tachikawashi ni okeru osen inryousui ni kansuru shitsumon shui-sho," *House of Representatives, Japan*, November 12, 1952, http://www.shugiin.go.jp/internet /itdb_shitsumona.nsf/html/shitsumon/a015004.htm (accessed June 14, 2019).

3. *Okinawa no beigun oyobi jieitai kichi* (Naha: Okinawa Prefectural Government, 2016), 110; "Shutoken ni mo ookatta beigunkichi—sono atochi kara mieru mono," *Yahoo! News*, October 4, 2017, https://news.yahoo.co.jp/feature/752 (accessed June 14, 2019).

4. Thomas R. H. Havens, *Fire across the Sea: The Vietnam War and Japan, 1965–1975* (Princeton, N.J.: Princeton University Press, 1987), 96. Havens describes Japan's support in the Vietnam War as follows: "American soldiers drank Kirin beer, chewed Lotte gum, and ate Chiba lettuce. American pilots dropped more than a billion propaganda leaflets, written in Vietnamese and printed in Kanagawa. Some of the injured received transfusions of Japanese blood, and those who died were sent home in polyethyline [*sic*] body bags made in Japan" (98).

5. Quoted in Havens, *Fire across the Sea*, 85.

6. "Kichi osen—kanagawa ken nai no omona ugoki," *Rimpeace*, http://www .rimpeace.or.jp/jrp/riku/pcb/kngwosen.html (accessed June 14, 2019).

7. Alvin Lee Young, *Investigations into Allegations of Herbicide Orange on Okinawa, Japan* (Cheyenne, Wyo.: A. L. Young Consulting, 2013), 9.

8. One of the US military's worst fires in Japan occurred in 1979, when a typhoon struck Camp Fuji, tipping over barrels of fuel that had been stored in the open; one marine was killed and another forty-one were seriously injured (Leon R. Yourtee and Gretchen Charles Greeson, *Castles in the Far East: The US Army Corps of Engineers Okinawa and Japan Districts, 1945–1990* [Japan: US Army, 1990]).

9. Sebastien Roblin, "How a Rocket Explosion Nearly Consumed a Nuclear-Powered Aircraft Carrier," *National Interest*, September 3, 2017, https://national interest.org/blog/the-buzz/how-rocket-explosion-nearly-consumed-nuclear-powered -22159 (accessed June 14, 2019).

10. Quoted in Havens, *Fire across the Sea*, 60.

11. Hans Kristensen, "Japan under the US Nuclear Umbrella," Nautilus Institute East Asia Nuclear Policy Project, 1999, https://nautilus.org/supporting -documents/japan-under-the-us-nuclear-umbrella (accessed June 14, 2019); Daniel Ellsberg, *The Doomsday Machine: Confessions of a Nuclear War Planner* (New York: Bloomsbury, 2017).

12. Ellsberg, *The Doomsday Machine*, 80, 117.

13. Kristensen, "Japan under the US Nuclear Umbrella," 9, quoted on 12.

14. One October 1958 US Navy report cited an agreement that allowed ships to bring nuclear weapons into Japanese ports without any notification. The claim was corroborated by later documents, including an April 1969 National Security Council (NSC) document and 1972 accounts from the commander in chief of the US Pacific Command (CINCPAC) and NSC (Kristensen, "Japan under the US Nuclear Umbrella," 36, 18, 3).

15. Kristensen, "Japan under the US Nuclear Umbrella," 27.

16. Kristensen, "Japan under the US Nuclear Umbrella," 38, 2.

17. Ellsberg, *The Doomsday Machine*, 70.

18. US Congress, House Committee on Armed Services, Readiness Subcommittee, Department of Defense Environmental Programs, Hearings. 102nd Congress, 1st Session March 21, April 17, 23, 24, and June 6, 1991 (Washington, D.C.: US Government Printing Office, 1991), iv, 788, 80–83.

19. Oshima Ken'ichi, "Kichi osen – kanagawa ken no omona ugoki," in *The State of the Environment in Asia 2005/2006*, ed. Takehisa Awaji and Shun'ichi Teranishi (Tokyo: Springer, 2005), 19–20. Also in the early 2000s, Kanagawa Prefecture cataloged large spills at Yokosuka Naval Base. In October 2003, there was a leak of thirty-four thousand liters of fuel oil from a tank from the anchored USS *Kitty Hawk*, followed by another spill of more than three thousand liters from the same ship in 2006 ("Kanagawa no beigun kichi," *Kanagawa Prefecture*, 2007, 137–39, http://www.pref.kanagawa.jp/docs/bz3/cnt/f417517/documents/22878.pdf [accessed June 14, 2019]).

20. "Kichi to akishima," Tokyoto Akishimashi, 2017, 21. "Kanagawa no beigun kichi," Kanagawa Prefecture, 2007, 138, accessed June 14, 2019, http://www.pref.kanagawa.jp/docs/bz3/cnt/f417517/documents/22878.pdf.

21. Among the Iwakuni incidents was one in June 2013, where more than one thousand liters of fuel spilled from an aircraft in an accident blamed on repair workers who had forgotten to replace parts of the airplane they had been fixing. Photographs show there was also a large spill of PCBs on the installation in May 2015, involving a leak into storm drains and off the base. Typical of the USMC's sloppy procedures, however, the released documents contained no actual details of the accident.

22. Another spill of firefighting foam occurred in October 2012; the accident report states that 4,164 liters was recovered, but it is unknown how much escaped into the environment. Atsugi's environmental problems may partly be due to the installation's lack of environmental staff. Between 2012 and 2014, the base lacked a division director for three months and a hazardous and PCB program compliance manager for four months.

23. The same water network was again contaminated in April 2015, by a leak of 380 liters of diesel, but the military delayed reporting it to local authorities. The largest incident occurred in October 2017; as many as three hundred thousand liters of sewage spilled from Misawa Air Base into Lake Anenuma, a source of drinking water. Civilian contractors took two weeks to report the leak to the US military, but even then, USFJ did not notify local authorities.

24. FOIA documents revealed the extent of the accident. The total volume of fuel released into the lake was approximately 2,271 liters, and at least one fish was found with fuel inside its body. Internal e-mails also show the wrong materials were deployed to soak up the spilled fuel, airmen did not receive sufficient training to perform the work, and they failed to spot fuel floating on the surface of the lake.

25. Nakayama Shoji et al., "Distributions of Perfluorooctanoic Acid (PFOA) and Perfluorooctane Sulfonate (PFOS) in Japan and Their Toxicities," *Environmental Sciences* 12, no. 6 (2005): 293–313.

26. "Kanagawa no beigun kichi," *Kanagawa Prefecture*, 137–38. In recent years, there have been other fires on US installations. According to internal documents I obtained from Yokosuka Naval Base, there was a fire at a hazardous waste storage facility there in August 2008, involving "various chemicals/hazardous waste"; however, according to the report, written on the day following the fire, Japanese authorities had not been notified about what had happened. Other reports I've obtained via FOIA have revealed details of a large fire at a hangar on Atsugi Air Base in November 2009. The fire burned thinners and other unspecified chemicals; extinguishing it created thousands of liters of wastewater likely contaminated with PFAS. According to the US Navy document, this wastewater was not cleared from the hangar until five days later. Three Japanese workers were injured in the blaze.

27. Erik Slavin, "Explosion at Sagami Depot Army Facility Near Tokyo Draws Criticism from Japanese," *Stars and Stripes*, August 25, 2015, https://www.stripes.com/news/pacific/explosion-at-sagami-depot-army-facility-near-tokyo-draws-criticism-from-japanese-1.364505 (accessed June 14, 2019).

28. Leon Cook, "No Foul Play Suspected in 2015 Explosion at Sagami Depot Near Tokyo," *Stars and Stripes*, November 8, 2016, https://www.stripes.com/news/no-foul-play-suspected-in-2015-explosion-at-sagami-depot-near-tokyo-1.438162 (accessed June 14, 2019).

29. In October 2015, 653 residents near Iwakuni were awarded a total of 558 million yen ($5.1 million). Likewise, in October 2017, the Tachikawa branch of the Tokyo District Court ordered the Japanese government to pay roughly 1,000 people living near Yokota Air Base approximately 610 million yen ($5.5 million), later raised to 760 million yen ($6.9 million) in June 2019 (Seth Robson and Kusumoto Hana, "Iwakuni Neighbors Win Noise Suit but Flights OK to Continue," *Stars and Stripes*, October 15, 2015, https://www.stripes.com/news/pacific/iwakuni-neighbors-win-noise-suit-but-flights-ok-to-continue-1.373374 [accessed June 14, 2019]; "Compensation Order Over Yokota Base Noise Upheld, Damages Raised," *Kyodo News*, June 6, 2019).

30. "Memorandum for Record/Subject: Shoriki Matsutaro," *Central Intelligence Agency*, January 27, 1953, https://www.cia.gov/library/readingroom/docs/SHORIKI%2C%20MATSUTARO%20%20%20VOL.%201_0014.pdf (accessed June 14, 2019).

31. For an English exploration of the role of the CIA in establishing NTV, see Richard Krooth, Morris Edelson, and Hiroshi Fukurai, *Nuclear Tsunami: The Japanese Government and America's Role in the Fukushima Disaster* (Lanham, Md.: Lexington, 2018), 3–38.

32. For a solid overview of CIA funding for the LDP, see Tim Weiner, *Legacy of Ashes: The History of the CIA* (New York: Doubleday, 2007), 133–140.

33. Eric Johnston, "Key Players Got Nuclear Ball Rolling," *Japan Times*, July 16, 2011, https://www.japantimes.co.jp/news/2011/07/16/national/key-players-got -nuclear-ball-rolling/ (accessed June 14, 2019).

34. "Dispatch Number FJBA-5662/Subject: Psych/Operational Relations with (redacted) PODAM," *Central Intelligence Agency*, July 5, 1955, 4, https://www .cia.gov/library/readingroom/docs/SHORIKI%2C%20MATSUTARO%20%20%20 VOL.%202_0013.pdf (accessed June 14, 2019).

35. SHORIKI, MATSUTARO VOL. 2_0013 3

36. Jeff Kingston, "Japan's Nuclear Village," *Asia-Pacific Journal* 10, issue 37, no. 1 (September 10, 2012): 1–23, https://apjjf.org/2012/10/37/Jeff-Kingston/3822 /article.html (accessed June 14, 2019).

37. In 2007, a 6.6 magnitude quake struck near the Kashiwazaki-Kariwa nuclear power station, Niigata Prefecture, which had been built to face tremors of a maximum 6.5 magnitude quake; the quake sparked a fire, and the plant was out of action until 2009.

38. Quoted in David Lochbaum, Edwin Lyman, Susan Q. Stranahan, and the Union of Concerned Scientists, *Fukushima: The Story of a Nuclear Disaster* (New York: New Press, 2014), 15.

39. "Tokyo Water 'Unfit for Babies' Due to High Radiation," *BBC*, March 23, 2011, https://www.bbc.com/news/world-asia-pacific-12825342 (accessed June 14, 2019); "Fukushima Water Headache: 1 Million Tons and Counting," *Asahi Shimbun*, March 19, 2019, http://www.asahi.com/ajw/articles/AJ201903190042.html (accessed June 14, 2019).

40. "Operation TOMODACHI – Japan Tsunami Relief Operations," Joint Task Force Civil Support, February 8, 2012, 48.

41. "Initial Impressions Report: Operation Tomodachi," Center for Army Lessons Learned, February 2012, 7; Matthew Burke and Elena Sugiyama, "Sasebo City Pushes Japanese to Dispose of Radioactive Waste on U.S. Bases," *Stars and Stripes,* September 30, 2011, https://www.stripes.com/news/sasebo-city-pushes-japanese -to-dispose-of-radioactive-waste-on-u-s-bases-1.156492 (accessed June 14, 2019).

42. According to USFJ, as of March 2018, solid low-level radioactive waste (LLRW) still remains at two of its installations. At Yokosuka Naval Base, there are forty-four pallets, and Yokota Air Base has three packages of the waste. USFJ declined to offer details on the actual weights involved.

43. "Operation Tomodachi," *Military Health System*, https://www.health.mil /Military-Health-Topics/Health-Readiness/Environmental-Exposures/Operation -Tomadachi (accessed June 14, 2019); Gregg Levine, "Seven Years on, Sailors Exposed to Fukushima Radiation Seek Their Day in Court," *Nation*, March 9, 2018, https://www.thenation.com/article/seven-years-on-sailors-exposed-to-fukushima -radiation-seek-their-day-in-court/ (accessed June 14, 2019).

44. Kageyama Yuri, "Sick US Sailors and Marines Who Blame Radiation Get Support from Japan's Ex-Leader," *Associated Press*, September 7, 2016, https:// www.navytimes.com/pay-benefits/military-benefits/2016/09/07/sick-u-s-sailors

-and-marines-who-blame-radiation-get-support-from-japan-s-ex-leader/ (accessed June 14, 2019).

45. Matthew Burke, "USS *Reagan* Sailors Not Exposed to High Radiation Levels in Japan, Report Finds," *Stars and Stripes*, July 17, 2014, https://www.stripes.com /news/uss-reagan-sailors-not-exposed-to-high-radiation-levels-in-japan-report -finds-1.293860 (accessed June 14, 2019).

46. Kageyama, "Sick US Sailors and Marines Who Blame Radiation Get Support from Japan's Ex-Leader"; Kristina Davis, "San Diego Judge Dismisses US Sailors' Fukushima Radiation Lawsuits, Rules Japan Has Jurisdiction," *San Diego Union-Tribune*, March 5, 2019, https://www.sandiegouniontribune.com/news/courts/sd -me-fukushima-lawsuits-dismissed-20190305-story.html (accessed June 14, 2019).

47. Quoted in Linda Pentz Gunter, "Injustice at Sea: The Irradiated Sailors of the USS *Reagan*," *Counterpunch*, March 7, 2018, https://www.counterpunch .org/2018/03/07/injustice-at-sea-the-irradiated-sailors-of-the-uss-reagan/ (accessed June 14, 2019).

48. Matthew Burke, "Sixteen US Ships That Aided in Operation Tomodachi Still Contaminated with Radiation," *Stars and Stripes*, March 13, 2016, https://www .stripes.com/news/special-reports/operation-tomodachi/16-us-ships-that-aided-in -operation-tomodachi-still-contaminated-with-radiation-1.399094 (accessed June 14, 2019).

49. Anna Fifield and Oda Yuki, "Japanese Nuclear Plant Just Recorded an Astronomical Radiation Level. Should We Be Worried?" *Washington Post*, February 8, 2017, https://www.washingtonpost.com/news/worldviews/wp/2017/02/08/japanese -nuclear-plant-just-recorded-an-astronomical-radiation-level-should-we-be-wor ried/ (accessed June 14, 2019).

50. "Fukushima Water Headache"; Takahashi Ryusei, "Eight Years after Triple Nuclear Meltdown, Fukushima No. 1's Water Woes Show No Signs of Ebbing," *Japan Times*, March 7, 2019, https://www.japantimes.co.jp/news/2019/03/07 /national/eight-years-triple-meltdown-fukushima-no-1s-water-woes-slow-recede/ (accessed June 14, 2019); "Radioactive Cesium above Legal Limit Detected in Fish Caught Off Fukushima," *Japan Times*, February 2, 2019, https://www.japantimes .co.jp/news/2019/02/02/national/science-health/limit-cesium-detected-fish-caught -off-fukushima/ (accessed June 14, 2019).

51. *On the Frontline of the Fukushima Nuclear Accident: Workers and Children* (Amsterdam: Greenpeace, March 2019), 37; "Disposing of Fukushima's Contaminated Soil," *NHK*, March 11, 2019, https://www3.nhk.or.jp/nhkworld/en/news /backstories/399/ (accessed June 14, 2019).

52. *On the Frontline of the Fukushima Nuclear Accident*, 7.

53. At White Beach, for instance, there were 590 visits between reversion and June 2019. In 2017, such submarines visited Sasebo Naval Base 26 times—the highest number ever recorded ("Genshiryoku gunkan no kikou," *Uruma City*, https:// www.city.uruma.lg.jp/shisei/167/908 [accessed June 14, 2019]).

54. Quoted in "A Mobile Nuclear Reactor Comes to Tokyo Bay," *Citizens' Nuclear Information Center*, January 12, 2006, http://www.cnic.jp/english/?p=970 (accessed June 14, 2019).

55. "Fact Sheet on US Nuclear-Powered Warship (NPW) Safety," *Japan Ministry of Foreign Affairs*, 2006, https://www.mofa.go.jp/region/n-america/us/security /fact0604.txt (accessed June 14, 2019).

56. "N-Warship Crisis Evacuation Starts after Radiation Doses Hit Twenty Times Higher Than N-Power Plant Crisis Manuals," *Japan Press Service*, May 13, 2014, http://www.japan-press.co.jp/s/news/?id=7183 (accessed June 14, 2019).

57. "Civilian Sub Passenger: 'I Thought the Sub Was Breaking Up,'" *CNN*, February 22, 2001, http://www.cnn.com/2001/US/02/22/japan.sub.02/index.html (accessed June 14, 2019); Gerry J. Gilmore, "Sub Skipper Reprimanded for *Ehime Maru* Incident," *American Forces Press Service*, April 25, 2001, http://www.defense .gov/news/newsarticle.aspx?id=44945 (accessed June 14, 2019); Thomas E. Ricks and Don Phillips, "Admiral Raises Questions in Sub Collision," *Washington Post*, February 15, 2001, https://www.washingtonpost.com/archive/politics/2001/02/15 /admiral-raises-questions-in-sub-collision/a6f5a054-1275-497f-a3af-e7e7959699c0/ (accessed June 14, 2019).

58. In January, the guided-missile cruiser USS *Antietam* ran aground in Tokyo Bay, spilling more than four thousand liters of hydraulic oil into the sea. In June, the destroyer USS *Fitzgerald* hit a container ship about one hundred kilometers southwest of Yokosuka, killing seven US sailors. The commanding officer and other officers were blamed for multiple safety failings and lost their commands. On August 21, the destroyer USS *John S. McCain* struck an oil tanker off the coast of Singapore. Ten sailors died. Again, the ship's senior officers were to blame—and the commander of the Seventh Fleet was fired as a result (Joe Sterling and Jason Hanna, "Plane Crash Latest in Spate of US Navy Accidents in Asia since January," *CNN*, November 22, 2017, https://edition.cnn.com/2017/08/21/politics/navy-ships -accidents/index.html [accessed June 14, 2019]).

59. Geoff Ziezulewicz, "This Sailor Brought Acid Aboard His Carrier," *Navy Times*, March 7, 2019, https://www.navytimes.com/news/your-navy/2019/03/07 /this-sailor-brought-acid-aboard-his-carrier/ (accessed June 14, 2019).

60. "US Nuclear Ship-Using Yokosuka Port Is Located Just Above Seismic Focal Zone," *Japan Press Service*, September 4, 2011, http://www.japan-press.co.jp/mod ules/news/index.php?id=2173 (accessed June 14, 2019).

## CHAPTER 8: TOXIC TERRITORIES: GUAM, THE COMMONWEALTH OF THE NORTHERN MARIANA ISLANDS, AND JOHNSTON ATOLL

1. Doug Mack, "The Strange Case of Puerto Rico," *Slate*, October 9, 2017, https://slate.com/news-and-politics/2017/10/the-insular-cases-the-racist-supreme-court-decisions-that-cemented-puerto-ricos-second-class-status.html (accessed June 14, 2019).

2. "U.S. Citizens Defend Democracy, Can't Vote for President," *Equally American*, http://www.equalrightsnow.org/u_s_citizens_defend_democracy_can_t_vote_for_president (accessed January 14, 2020).

3. Valeria Pelet, "Puerto Rico's Invisible Health Crisis," *The Atlantic*, September 3, 2016, https://www.theatlantic.com/politics/archive/2016/09/vieques-invisible-health-crisis/498428/ (accessed June 14, 2019).

4. Quoted in Doug Herman, "A Brief, Five Hundred-Year History of Guam," *Smithsonian.com*, August 15, 2017, https://www.smithsonianmag.com/smithsonian-institution/brief-500-year-history-guam-180964508/ (accessed June 14, 2019).

5. Jayne Aaron, *Regional Cold War History for Department of Defense Installations in Guam and the Northern Mariana Islands* (Washington, D.C.: Department of Defense Legacy Resource Management Program, 2011), 16, 72.

6. Quoted in Donald F. McHenry, *Micronesia: Trust Betrayed* (Washington, D.C.: Carnegie Endowment for International Peace, 1975), 66.

7. Quoted in Aaron, *Regional Cold War History for Department of Defense Installations in Guam and the Northern Mariana Islands*, 29.

8. David Vine, *Base Nation: How US Military Bases Abroad Harm America and the World* (New York: Metropolitan Books, 2015), 145; McHenry, *Micronesia*, 7.

9. Attempts to eliminate Guam's brown tree snakes have been unsuccessful, leading to increasingly desperate countermeasures. In operations reminiscent of something devised by Unit 731, in 2013 the Department of Agriculture began injecting dead mice with acetaminophen, the active ingredient of Tylenol, which cures human headaches but can be lethal to snakes. The rodents' drugged carcasses were then strapped into paper parachutes and dropped into the snakes' habitat. In March 2019, six thousand capsules of acetaminophen-spiked mice fetuses were scattered throughout the jungles on Andersen Air Force Base, with results that have proven inconclusive (Mark Memmott, "Dead Mice Update: Tiny Assassins Dropped on Guam Again," *NPR*, December 3, 2013, https://www.npr.org/sections/thetwo-way/2013/12/03/248386912/dead-mice-update-tiny-assassins-dropped-on-guam-again [accessed June 14, 2019]; "Brown Tree Snake Aerial Bait System Used Up North," *KUAM News*, March 22, 2019, https://www.kuam.com/story/40177182/2019/03/22/brown-tree-snake-aerial-bait-system-used-up-north [accessed June 14, 2019]).

10. Aaron, *Regional Cold War History for Department of Defense Installations in Guam and the Northern Mariana Islands*, 62.

11. McHenry, *Micronesia*, 13, 28.

12. Aaron, *Regional Cold War History for Department of Defense Installations in Guam and the Northern Mariana Islands*, 89.

13. "History of the Custody and Deployment of Nuclear Weapons: July 1945 through September 1977, Appendix B," *Office of the Secretary of Defense*, 1978,

https://nsarchive2.gwu.edu//news/19991020/04-47.htm (accessed June 14, 2019); Aaron, *Regional Cold War History for Department of Defense Installations in Guam and the Northern Mariana Islands*, 75–76.

14. *Assessment of the Scientific Information for the Radiation Exposure Screening and Education Program* (Washington, D.C.: National Academies Press, 2005), 365; Jonathan M. Weisgall, *Operation Crossroads: The Atomic Tests at Bikini Atoll* (Annapolis, Md.: Naval Institute Press, 1994), 322.

15. Mar-Vic Cagurangan, "War Vet to Apologize for Guam Radiation Cover-Up," *Pacific Islands Report*, November 4, 2005, http://www.pireport.org/articles /2005/11/04/war-vet-apologize-guam-radiation-coverup (accessed June 14, 2019).

16. *Assessment of the Scientific Information for the Radiation Exposure Screening and Education Program*, 200.

17. "Radiation Compensation Plan Considered," *KUAM News*, April 23, 2015, https://www.kuam.com/story/28876875/2015/04/23/radiation-compensation-plan -considered (accessed June 14, 2019).

18. In 2018, Guam senator Therese Terlaje stated, "It's an unsettling truth that our family members were exposed to radiation from US nuclear testing, and suffer from cancer and other illnesses. . . . Since the findings in 2005, Guam's inclusion in RECA is overdue. Compassion, medical care, and justice for the people of Guam suffering from cancer is equally warranted" ("Pacific Association for Radiation Survivors Continues Push for Radiation Compensation," *Pacific Daily News*, August 1, 2018, https://www.guampdn.com/story/news/2018/08/01/pacific-associa tion-radiation-survivors-continues-push-compensation/877399002/ [accessed June 14, 2019]).

19. David B. Kaplan, "When Incidents Are Accidents: The Silent Saga of the Nuclear Navy," *Oceans Magazine*, August 1983, http://oc.itgo.com/kitsap/nuclear /clymer.htm (accessed June 14, 2019); "Navy Maintains USS *Houston* Leak Poses No Risk," *KUAM News*, August 8, 2008, https://www.kuam.com/story/11077472 /navy-maintains-uss-houston-leak-poses-no-risk (accessed June 14, 2019); Clynt Ridgell, "Navy Officials Say No Radiation Leaked into Guam Environment from Nuclear-Powered Sub," *Pacific News Center*, November 24, 2014, https://pnc guam.com/navy-officials-say-no-radiation-leaked-into-guam-environment-from -nuclear-powered-sub/ (accessed January 14, 2020).

20. Untitled TV commercial, USAF, aired March 15, 2019.

21. "B-2 Mission Commander Recounts February Crash on Guam Runway," *Stars and Stripes*, August 15, 2008, https://www.stripes.com/news/b-2-mission -commander-recounts-february-crash-on-guam-runway-1.81994 (accessed June 14, 2019); "B-2 Fire at AAFB Back in February of 2010 Was 'Horrific,' Not 'Minor,'" *Pacific News Center*, August 31, 2011, https://pncguam.com/wright-patterseon/ (accessed June 14, 2019).

22. Letter from Philip W. Grone, principal assistant deputy undersecretary of defense (Installations and Environment) to Congressman Lane Evans, September 23, 2003.

23. Jon Mitchell, "Poisons in the Pacific: Guam, Okinawa, and Agent Orange," *Japan Times*, August 7, 2012, https://www.japantimes.co.jp/community/2012/08/07/issues/poisons-in-the-pacific-guam-okinawa-and-agent-orange/ (accessed June 14, 2019).

24. Quoted in Mitchell, "Poisons in the Pacific: Guam, Okinawa, and Agent Orange."

25. Quoted in Mitchell, "Poisons in the Pacific: Guam, Okinawa, and Agent Orange."

26. Jonathan K. Noel, Sara Namazi, and Robert L. Haddock, "Disparities in Infant Mortality Due to Congenital Anomalies on Guam," *Hawaii Journal of Medicine and Public Health*, 74, no. 12 (December 2015): 397–402.

27. Haidee V Eugenio, "Some Veterans Got Agent Orange Benefits, Many More Still Waiting," *Pacific Daily News*, August 4, 2014, https://www.guampdn.com/story/news/2018/08/04/some-veterans-got-agent-orange-benefits-many-more-still-waiting/858650002/ (accessed June 14, 2019).

28. Haidee V Eugenio, "Bill Seeks Help for Fifty-Two Thousand Exposed to Toxic Herbicides on Guam, Region," *Pacific Daily News*, March 14, 2019, https://www.guampdn.com/story/news/local/2019/03/13/thousands-exposed-toxic-herbicides-guam/3149578002/ (accessed June 14, 2019).

29. *Hazardous Waste: Abandoned Disposal Sites May Be Affecting Guam's Water Supply* (Washington, D.C.: General Accounting Office, 1987), 15.

30. *Environmental Cleanup: Better Communication Needed for Dealing with Formerly Used Defense Sites in Guam* (Washington, D.C.: General Accounting Office, 2002), 7.

31. In Urunao, Dededo Village, the USAF dumped aircraft parts, barrels, and UXO, notably one 680-kilogram bomb; at times, the military tried using napalm to burn the scrap, but this only exacerbated the problem. Contaminants later detected included arsenic, heavy metals, PCBs, and dioxins; the volume of waste was estimated at approximately eighteen hundred truckloads. Meanwhile, the navy dumped such refuse as paint thinners and battery casings in a similar way over cliffs at Orote Point.

32. Neil Pang, "GovGuam's $160M Suit against Navy Raises Dumping of Military Waste," *Guam Daily Post*, March 8, 2017, https://www.postguam.com/news/local/govguam-s-m-suit-against-navy-raises-dumping-of-military/article_142686d0-02d8-11e7-b5bc-ff41f549736a.html (accessed June 14, 2019).

33. Mindy Aguon, "Three Hundred Twenty Tons of PCBs Removed from Agana Springs," *KUAM News*, April 22, 2012, https://www.kuam.com/story/17654126/320-tons-of-pcbs-removed-from-agana-springs (accessed June 14, 2019); "US EPA Cleans Up PCB Contamination in Agana Springs," *Pacific News Center*, December 16, 2011, https://pacificnewscenter.com/us-epa-cleans-up-pcbs-contamination-in-agana-springs/ (accessed June 14, 2019).

34. Robert L. Haddock, Grazyna Badowski, and Renata Bordallo, "Mortality Following Polychlorinated Biphenyl (PCB) Contamination of a Guam Village," *Hawaii Medical Journal* 70, no. 11, supplement 2 (November 2011): 40–42.

35. Haidee V Eugenio, "More PCB Contamination Investigation in Cocos Lagoon," *Pacific Daily News*, February 13, 2019, https://www.guampdn.com/story /news/local/2019/02/13/more-pcb-contamination-investigation-cocos-lagoon /2845375002/ (accessed June 14, 2019).

36. Haddock et al., "Mortality," 42.

37. John I Borja, "Twenty-Five Years Later, Andersen Air Force Base Still Cleaning Up Contamination, EPA Says," *Pacific Daily News*, October 15, 2017, https://www.guampdn.com/story/news/2017/10/15/25-years-later-andersen-air -force-base-still-cleaning-up-contamination-epa-says/764211001/ (accessed June 14, 2019); interview with Guam EPA, March 12, 2019; interview with Joint Region Marianas, March 15, 2019.

38. John O'Connor, "Feds to Fund Study of Toxic Chemicals," *Guam Post*, November 26, 2017, https://www.postguam.com/news/local/feds-to-fund-study -of-toxic-chemicals/article_209fc08a-d0b8-11e7-926c-5f575fea433a.html (accessed June 14, 2019); "Our View: Whale Deaths Should Prompt Us to Ask Navy for More Information about Sonar," *Pacific Daily News*, March 4, 2019, https://www .guampdn.com/story/opinion/editorials/2019/03/04/we-should-ask-more-questions -whale-deaths-and-sonar-our-view/3052570002/ (accessed June 14, 2019).

39. LisaLinda Natividad and Victoria Lola Leon-Guerrero, "The Explosive Growth of US Military Power on Guam Confronts People Power: Experience of an Island People under Spanish, Japanese, and American Colonial Rule," *Asia-Pacific Journal* 49, issue 3, no. 10 (December 6, 2010): 1–16, https://apjjf.org/-LisaLinda -Natividad/3454/article.html (accessed June 14, 2019).

40. Quoted in "Looking for Friendly Overseas Base, Pentagon Finds It Already Has One," *New York Times*, April 7, 2004, https://www.nytimes.com/2004/04/07 /us/looking-for-friendly-overseas-base-pentagon-finds-it-already-has-one.html (accessed June 14, 2019).

41. Natividad and Leon-Guerrero, "The Explosive Growth of US Military Power on Guam Confronts People Power."

42. Interview with Senator Therese Terlaje, March 14, 2019.

43. "Commonwealth of the Northern Mariana Islands Joint Military Training Environmental Impact Statement/Overseas Environmental Impact Statement (Draft)," US Marine Corps Forces Pacific, 2015, 610–12, 606.

44. Gary R. W. Denton, Carmen A. Emborski, April A. B. Hachero, Ray S. Masga, and John A. Starmer, "Impact of WWII Dump Sites on Saipan (CNMI): Heavy Metal Status of Soils and Sediments," *Environmental Science and Pollution Research* 23, no. 11 (June 2016): 11,339–48. 347.

45. Roger B. Jeans, *The CIA and Third Force Movements in China during the Early Cold War: The Great American Dream* (Lanham, Md.: Lexington, 2018), 127–28; William H. Stewart, "Cold War Covert Activities on Saipan, Elsewhere

in the Region (Part One)," *Saipan Tribune*, December 21, 2004, https://www
.saipantribune.com/index.php/a1c6be2c-1dfb-11e4-aedf-250bc8c9958e/ (accessed
June 14, 2019).

46. Jeans, *The CIA and Third Force Movements in China during the Early Cold
War*, 127–28.

47. "Naval Technical Training Unit (NTTU)," *Pacific Worlds CNMI-Tanapag*,
http://www.pacificworlds.com/cnmi/memories/memory3.cfm (accessed June 14,
2019).

48. Anita Hofschneider, "Can These Islands Survive America's Military Pivot
to Asia?" *Civil Beat*, December 2012, https://www.civilbeat.org/2016/12/can-these
-islands-survive-americas-military-pivot-to-asia/ (accessed June 14, 2019).

49. Between October 1999 and April 2001, the *Saipan Tribune* published a series of articles tracking the contamination at Tanapag and the failure of authorities
to remediate it. Key articles during this period include the following: "PCB and
Dioxin Contamination: Tanapag Villagers Seek Justice," *Saipan Tribune*, November
3, 1999, https://www.saipantribune.com/index.php/95175545-1dfb-11e4-aedf
-250bc8c9958e/ (accessed June 14, 2019); "Cemetery Off-Limits," *Saipan Tribune*,
October 28, 1999, https://www.saipantribune.com/index.php/b1fe6aa5-1dfb-11e4
-aedf-250bc8c9958e/ (accessed June 14, 2019); "DPH Warns against Tanapag Land
Crabs," *Saipan Tribune*, October 3, 2000, https://www.saipantribune.com/index
.php/9641db09-1dfb-11e4-aedf-250bc8c9958e/ (accessed June 14, 2019); "EPA
Assails US Army Corps for Slow Action," *Saipan Tribune*, April 26, 2000, https://
www.saipantribune.com/index.php/95c2692c-1dfb-11e4-aedf-250bc8c9958e/ (accessed June 14, 2019); "The PCB Fiasco," *Saipan Tribune*, August 21, 2000, https://
www.saipantribune.com/index.php/9626ad67-1dfb-11e4-aedf-250bc8c9958e/ (accessed June 14, 2019); "PCB Warfare among Agencies," *Saipan Tribune*, December 15, 2000, https://www.saipantribune.com/index.php/968006a1-1dfb-11e4-aedf
-250bc8c9958e/ (accessed June 14, 2019); "AGO Says Feds Mishandled PCB
Problem," *Saipan Tribune*, August 4, 2000, https://www.saipantribune.com/index
.php/961b29e2-1dfb-11e4-aedf-250bc8c9958e/ (accessed June 14, 2019); "Dioxin
and PCB Contamination in Tanapag: 'Our People Are Sick and Dying,'" *Saipan
Tribune*, November 1, 1999, https://www.saipantribune.com/index.php/95171f89
-1dfb-11e4-aedf-250bc8c9958e/ (accessed June 14, 2019); "Gov't Still Undecided
on PCB Lawsuit," *Saipan Tribune*, November 7, 2000, https://www.saipantribune
.com/index.php/965433cf-1dfb-11e4-aedf-250bc8c9958e/ (accessed June 14, 2019);
"New Study on Toxic Wastes in Tanapag," *Saipan Tribune*, April 26, 2001, https://
www.saipantribune.com/index.php/9736edea-1dfb-11e4-aedf-250bc8c9958e/ (accessed June 14, 2019).

50. "Feds to Test Saipan for Agent Orange Contamination," *Hawaii Public
Radio*, April 26, 2000, http://www.pireport.org/articles/2000/04/28/feds-test-saipan
-agent-orange-contamination (accessed June 14, 2019); "Massive Cleanup of Contaminated Site on Capitol Hill," *Saipan Tribune*, February 7, 2007, https://www
.saipantribune.com/index.php/be239369-1dfb-11e4-aedf-250bc8c9958e/ (accessed

June 14, 2019); "Army Corps Finishes I-Denni Excavation," *Saipan Tribune*, May 21, 2012, https://www.saipantribune.com/index.php/bf8e41e2-1dfb-11e4 -aedf-250bc8c9958e/ (accessed June 14, 2019). In 2012, twelve thousand UXO were removed from land slated for housing construction in Marpi village. On neighboring Tinian Island, contamination from UXO, chromium, and iron was found on a mortar range closed in 1994. Also, on Tinian's 118-hectare Masalog Ridge Area Site, postwar storage of ordnance had left behind UXO and potentially other substances, which, according to an official investigation in 2014, presented a "significant health hazard to general public" (Haidee V. Eugenio, "$4M to Clean Up Three Saipan Sites," *Saipan Tribune*, April 7, 2014, https://www.saipantribune .com/index.php/4m-clean-3-saipan-sites/ [accessed June 14, 2019]; "Commonwealth of the Northern Mariana Islands Joint Military Training Environmental Impact Statement/Overseas Environmental Impact Statement (Draft)," US Marine Corps Forces Pacific, 606, 612).

51. Anita Hofschneider, "Missing Data Plagues Military Training Plans in the Marianas," *Honolulu Civil Beat*, December 2016, https://www.civilbeat .org/2016/12/what-the-military-isnt-saying-about-its-training-plans-in-the-mari anas/ (accessed June 14, 2019); Anita Hofschneider, "The Fight to Save Pagan Island from US Bombs," *Honolulu Civil Beat*, December 2016, https://www.civil beat.org/2016/12/the-fight-to-save-pagan-island-from-us-bombs/ (accessed June 14, 2019).

52. Peter J. Perez, "The Military Takeover of the Marianas," *Saipan Tribune*, September 25, 2015, https://www.saipantribune.com/index.php/the-military-take over-of-the-marianas/ (accessed June 14, 2019).

53. Annie Jacobsen, *Area 51: An Uncensored History of America's Top-Secret Military Base* (New York: Little, Brown and Company, 2012), 170–72.

54. "Operation Dominic I—1962," Defense Nuclear Agency, 1983, 238, 250, 253.

55. "Operation Dominic I—1962," 106.

56. "Operation Dominic I—1962," 234.

57. "Operation Dominic I—1962," 363, 238.

58. "Cleaning Up Johnston Atoll," *APSNet Special Reports*, November 25, 2005, https://nautilus.org/apsnet/cleaning-up-johnston-atoll/ (accessed June 14, 2019).

59. R. J. Ritter (ed.), "Newsletter of National Association of Atomic Veterans, Inc.," March 2010.

60. William A. Buckingham, *Operation Ranch Hand: The Air Force and Herbicides in Southeast Asia, 1961–1971* (Washington, D.C.: Office of Air Force History, 1982), 188; Fred A. Wilcox, *Waiting for an Army to Die: The Tragedy of Agent Orange* (Santa Ana, Calif.: Seven Locks Press, 1989), 121; Buckingham, *Operation Ranch Hand*, 188.

61. "Preliminary Public Health, Environmental Risk, and Data Requirements Assessment for the Herbicide Orange Storage Site at Johnston Island," *Air Force Armstrong Laboratory*, 1991, https://web.archive.org/web/20130421032235/http:// www.guamagentorange.info/yahoo_site_admin/assets/docs/Johnston_Atoll_His

tory.261114404.pdf (accessed June 14, 2019); "Final Phase II Environmental Baseline Survey Report, Johnston Atoll," Earth Tech, 2005, B-22; "ATSDR to Hold Open House about the Naval Construction Battalion Center Site in Gulfport, Miss.," *Agency for Toxic Substances and Disease Registry*, August 12, 2002, https://www.atsdr.cdc.gov/news/displaynews.asp?PRid=1955 (accessed June 14, 2019).

62. "At-Sea Incineration of Herbicide Orange Onboard the *M/T Vulcanus*," US Air Force and Environmental Protection Agency, April 1978.

63. Buckingham, *Operation Ranch Hand*, 189.

64. National Research Council, *Evaluation of Chemical Events at Army Chemical Agent Disposal Facilities* (Washington, D.C.: National Academies Press, 2002), 75–76, https://doi.org/10.17226/10574; Donovan Webster, *Aftermath: The Remnants of War* (New York: Pantheon, 1996), Loc. 3532.

65. National Research Council, *Evaluation of Chemical Events at Army Chemical Agent Disposal Facilities*, 104, 27.

66. National Research Council, *Evaluation of Chemical Events at Army Chemical Agent Disposal Facilities*, 24, 9.

67. "Johnston Atoll Decommissioning," *Parsons Corporation*, https://www.parsons.com/construction/environmental-radiation-and-restoration/projects/johnston-atoll.asp (accessed June 14, 2019); "Final Phase II Environmental Baseline Survey Report, Johnston Atoll," B-31, B-33.

68. Lisa Kerr Lobel and Phillip S. Lobel, "Contaminants from Fishes from Johnston Island," Proceedings of the 11th International Coral Reef Symposium, Ft. Lauderdale, Florida, July 7–11, 2008; Lisa Kerr Lobel, "Toxic Caviar: Using Fish Embryos to Monitor Contaminant Impacts," in *Diving for Science, Proceedings of the American Academy of Underwater Sciences 30th Symposium*, ed. Neal Pollock (Dauphin Island, Ala.: American Academy for Underwater Sciences, 2011), 33–39.

69. "Herbicide Tests and Storage Outside the US," *Department of Veterans Affairs*, https://www.publichealth.va.gov/exposures/agentorange/locations/tests-storage/outside-vietnam.asp (accessed April 15, 2019).

70. Bob Drogin, "Erasing a Chemical Arsenal," *Los Angeles Times*, August 23, 2009, https://www.latimes.com/archives/la-xpm-2009-aug-23-na-chemical23-story.html (accessed April 15, 2019).

## CHAPTER 9: TOWARD ENVIRONMENTAL JUSTICE

1. For an exploration of the life and legacy of Angel Santos, see Michael Lujan Bevacqua, "Interpretive Essay: Angel L. G. Santos," *Guampedia*, https://www.guampedia.com/interpretive-essay-angel-l-g-santos/ (accessed June 14, 2019).

2. Chris Gelardi and Sophia Perez, "'This Isn't Your Island': Why Northern Mariana Islanders Are Facing Down the US Military," *Nation*, June 12, 2019, https://www.thenation.com/article/northern-mariana-islands-military-bases-tinian/ (accessed June 14, 2019).

3. Mukai Daisuke, "Thesis by Unit 731 Member an Issue as Human Testing Feared," *Asahi Shimbun*, August 29, 2018, http://www.asahi.com/ajw/articles /AJ201808290006.html (accessed June 14, 2019).

4. "Agent Orange: Actions Needed to Improve Accuracy and Communication of Information on Testing and Storage Locations," *Government Accountability Office*, November 2018, 56, https://www.gao.gov/assets/700/695490.pdf (accessed June 14, 2019).

5. "'Chouju no ken' shokku futatabi/Okinawa heikin jumyou josei 7-i, dansei 36-i ni koutai," *Okinawa Times*, December 14, 2017, https://www.okinawatimes.co.jp/ articles/-/183880 (accessed June 14, 2019).

# BIBLIOGRAPHY

Aaron, Jayne. *Regional Cold War History for Department of Defense Installations in Guam and the Northern Mariana Islands.* Washington, D.C.: Department of Defense Legacy Resource Management Program, 2011.

Ahagon, Shoko. *The Island Where People Live,* trans. C. Harold Rickard. Hong Kong: Christian Conference of Asia Communications, 1989.

*Assessment of the Scientific Information for the Radiation Exposure Screening and Education Program.* Washington, D.C.: National Academies Press, 2005.

Barenblatt, Daniel. *A Plague upon Humanity.* New York: HarperCollins, 2004.

Bix, Herbert B. *Hirohito and the Making of Modern Japan.* New York: Harper-Collins, 2000.

Braw, Monica. *The Atomic Bomb Suppressed: American Censorship in Occupied Japan.* New York: M. E. Sharpe, 1991.

Buckingham, William A. *Operation Ranch Hand: The Air Force and Herbicides in Southeast Asia, 1961–1971.* Washington, D.C.: Office of Air Force History, 1982.

Chakma, Suhas, and Marianne Jensen (eds.). *Racism against Indigenous Peoples.* Copenhagen: International Work Group for Indigenous Affairs, 2001.

Chuugoku Shimbun. *Dokugasu no Shima: Okunoshima – Akumu no Kizuato.* Hiroshima: Chuugoku Shimbunsha, 1996.

Cook, Haruko Taya, and Theodore F. Cook. *Japan at War: An Oral History.* New York: New Press, 1992.

Cunliffe, William H. *Select Documents on Japanese War Crimes and Japanese Biological Warfare, 1934–2006.* Washington, D.C.: National Archives and Records Administration, 2007.

Dibblin, Jane. *Day of Two Suns: U.S. Nuclear Testing and the Pacific Islanders.* New York: New Amsterdam Books, 1990.

Dower, John W. *Embracing Defeat: Japan in the Wake of World War II.* New York: W. W. Norton & Company, 1999.

Dower, John W. *Japan in War and Peace: Essays on History, Culture, and Race.* London: HarperCollins, 1995.

Durant, Robert F. *The Greening of the U.S. Military: Environmental Policy, National Security, and Organizational Change.* Washington, D.C.: Georgetown University Press, 2007.

Ellsberg, Daniel. *The Doomsday Machine: Confessions of a Nuclear War Planner.* New York: Bloomsbury, 2017.

Glacken, Clarence J. *The Great Loochoo: A Study of Okinawan Village Life.* Berkeley: University of California Press, 1955.

Gonzalez, Ricardo M., and Mark D. Merlin. "Environmental Impacts of Nuclear Testing in Remote Oceania, 1946—1996." In *Environmental Histories of the Cold War,* ed. J. R McNeill and Corinna R. Unger (New York: Cambridge University Press, 2010), 167–202.

Griffiths, Philip Jones. *Agent Orange: "Collateral Damage" in Vietnam.* London: Trolley Limited, 2003.

Grunden, Walter E. "No Retaliation in Kind: Japanese Chemical Warfare Policy in World War II." In *One Hundred Years of Chemical Warfare: Research, Deployment, Consequences,* ed. Bretislav Friedrich, Dieter Hoffmann, Jurgen Renn, Florian Schmaltz, and Martin Wolf (Cham: Springer Nature, 2017), 259–71.

Harris, Sheldon H. *Factories of Death: Japanese Biological Warfare, 1932–1945, and the American Cover-up.* London: Routledge, 1994.

———. *Factories of Death: Japanese Biological Warfare, 1932–1945, and the American Cover-up,* 2nd ed. London: Routledge, 2002.

Havens, Thomas R. H. *Fire across the Sea: The Vietnam War and Japan, 1965–1975.* Princeton, N.J.: Princeton University Press, 1987.

Hay, Alastair. *The Chemical Scythe: Lessons of 2,4,5-T and Dioxin.* New York: Springer US, 1982.

Hiroshima Peace Memorial Museum. *The Spirit of Hiroshima.* Hiroshima: Hiroshima Peace Memorial Museum, 1999.

Horiba, Kiyoko. *Hiroshima and Nagasaki Censored.* Tokyo: International Peace Research Institute, Meiji Gakuin University, 2018.

Jacobsen, Annie. *Area 51: An Uncensored History of America's Top-Secret Military Base.* New York: Little, Brown and Company, 2012.

Jeans, Roger B. *The CIA and Third Force Movements in China during the Early Cold War: The Great American Dream.* Lanham, Md.: Lexington, 2018.

Josephson, Paul. "War on Nature as Part of the Cold War: The Strategic and Ideological Roots of Environmental Degradation in the Soviet Union," in *Environmental Histories of the Cold War,* ed. J. R. McNeill and Corinna R. Unger. New York: Cambridge University Press, 2010.

Krooth, Richard, Morris Edelson and Hiroshi Fukurai. *Nuclear Tsunami: The Japanese Government and America's Role in the Fukushima Disaster.* Lanham, Md.: Lexington, 2018.

Lifton, Robert Jay, and Greg Mitchell. *Hiroshima in America: Fifty Years of Denial.* New York: G. P. Putnam's Sons, 1995.

Lochbaum, David, Edwin Lyman, Susan Q. Stranahan, and the Union of Concerned Scientists. *Fukushima: The Story of a Nuclear Disaster.* New York: New Press, 2014.

McHenry, Donald F. *Micronesia: Trust Betrayed.* Washington, D.C.: Carnegie Endowment for International Peace, 1975.

McCormack, Gavan and Satoko Oka Norimatsu. *Resistant Islands: Okinawa Confronts Japan and the United States.* Lanham, Md.: Rowman and Littlefield, 2012.

Mitchell, Greg. *Atomic Cover-up: Two U.S. Soldiers, Hiroshima and Nagasaki, and the Greatest Movie Never Made.* New York: Sinclair Books, 2011.

Morris, M. D. *Okinawa: A Tiger by the Tail.* New York: Hawthorn, 1968.

NARA staff, "Documentary Evidence and Studies of Japanese War Crimes: An Interim Assessment." In *Researching Japanese War Crimes Records: Introductory Essays*, ed. Edward Drea, Greg Bradsher, Robert Hanyok, James Lide, Michael Petersen, and Daqing Yang (Washington, D.C.: Nazi War Crimes and Japanese Imperial Government Records Interagency Working Group, 2006), 79–111.

Oishi, Matashichi. *The Day the Sun Rose in the West: Bikini, the Lucky Dragon, and I.* Honolulu: University of Hawai'i Press, 2011.

Okinawa ken kyouiku iinkai (eds.). *Okinawa Ken Shi.* Naha: Okinawa ken kyouiku iinkai, 1974.

Oshima Ken'ichi, "Kichi osen – kanagawa ken no omona ugoki." In *The State of the Environment in Asia 2005/2006*, ed. Takehisa Awaji and Shun'ichi Teranishi. Tokyo: Springer, 2005.

Ota, Masahide. *Essays on Okinawa Problems.* Gushikawa: Yui Shuppan Company, 2000.

Rabson, Steve. *The Okinawan Diaspora in Japan: Crossing the Borders Within.* Honolulu: University of Hawai'i Press, 2012.

Rhodes, Richard. *The Making of the Atomic Bomb: The 25th Anniversary Edition.* New York: Simon and Schuster Paperbacks, 2012.

Rotter, Andrew J. *Hiroshima: The World's Bomb.* New York: Oxford University Press, 2008.

Ryukyu Shimpo. *Descent into Hell: Civilian Memories of the Battle of Okinawa.* Portland, Maine: MerwinAsia, 2014.

Sanders, Barry. *The Green Zone: The Environmental Costs of Militarism.* Oakland, Calif.: AK Press, 2009.

Sarantakes, Nicholas Evan. *Keystone: The American Occupation of Okinawa and US–Japanese Relations.* College Station: Texas A&M University Press, 2000.

Shulman, Seth. *The Threat at Home: Confronting the Toxic Legacy of the U.S. Military.* Boston: Beacon, 1992.

Southard, Susan. *Nagasaki: Life after Nuclear War.* New York: Penguin, 2016.

Stone, Oliver, and Peter Kuznick. *The Untold History of the United States.* New York: Simon and Schuster, 2012.

Tucker, Jonathan B. *War of Nerves: Chemical Warfare from World War I to Al-Qaeda.* New York: Anchor Books, 2006.

Ui, Jun (ed.). *Industrial Pollution in Japan.* Tokyo: United Nations University Press, 1992.

Vine, David. *Base Nation: How U.S. Military Bases Abroad Harm America and the World.* New York: Metropolitan Books, 2015.

Waugh, Charles, and Huy Lien. *Family of Fallen Leaves: Stories of Agent Orange by Vietnamese Writers.* Athens: University of Georgia Press, 2010.

Webster, Donovan. *Aftermath: The Remnants of War.* New York: Pantheon, 1996.

Weiner, Tim. *Legacy of Ashes: The History of the CIA.* New York: Doubleday, 2007.

Weisberg, Barry. *Ecocide in Indochina: The Ecology of War.* San Francisco, Calif.: Canfield Press, 1970.

Weisgall, Jonathan M. *Operation Crossroads: The Atomic Tests at Bikini Atoll.* Annapolis, Md.: Naval Institute Press, 1994.

Weller, George. *First into Nagasaki: The Censored Eyewitness Dispatches on Post-Atomic Japan and Its Prisoners of War.* New York: Crown, 2006.

Wilcox, Fred A. *Scorched Earth: Legacies of Chemical Warfare in Vietnam.* New York: Seven Stories Press, 2011.

———. *Waiting for an Army to Die: The Tragedy of Agent Orange.* Santa Ana, Calif.: Seven Locks Press, 1989.

Williams, Peter, and David Wallace. *Unit 731: The Japanese Army's Secret of Secrets.* London: Hodder and Stoughton, 1989.

Yang, Daqing. "Documentary Evidence and Studies of Japanese War Crimes: An Interim Assessment." In *Researching Japanese War Crimes Records: Introductory Essays,* ed. Edward Drea, Greg Bradsher, Robert Hanyok, James Lide, Michael Petersen, and Daqing Yang (Washington, D.C.: Nazi War Crimes and Japanese Imperial Government Records Interagency Working Group, 2006), 21–56.

Yoshida, Kensei. *Democracy Betrayed: Okinawa under US Occupation.* Bellingham: Western Washington University, 2001.

Zierler, David. *The Invention of Ecocide: Agent Orange, Vietnam, and the Scientists Who Changed the Way We Think about the Environment.* Athens: University of Georgia Press, 2011.

# ACKNOWLEDGMENTS

In the summer of 2010, I traveled to the Yanbaru jungles of Northern Okinawa to report for the *Japan Times* about the impact of military operations on the island's environment. For the first time, I heard how US troops there in the 1960s had sprayed Agent Orange to clear vegetation and built mock Vietnamese villages to stage war games. Residents described early deaths among base workers, worried their land was still contaminated, and feared for their children's health—but the Japanese government had done nothing to help. I promised them I'd investigate, so after returning to Tokyo, I started reaching out to US service members who had served on Okinawa during the Vietnam War. These veterans recalled offloading orange-banded barrels of defoliants and spraying them throughout the island; many were ill and their children were sick, and the VA was ignoring their plight.

Hearing these parallel experiences from Okinawans and Americans—their illnesses, fears for the next generation, and abandonment by the authorities—triggered the start of this investigation into military contamination throughout the Pacific region. During my decade-long journey I've been guided by hundreds of people without whose support, advice, and show of faith this book would not exist.

On Okinawa, I offer my heartfelt thanks to Ashimine Gentatsu and Yukine and Isa Ikuko for their knowledge of the Yanbaru; Professor Sato Manabu of Okinawa International University for his insights into US–Japan relations; Kuniyoshi Nagahiro for his tales of journalism in the prereversion era; interpreter Oshiro Nariko for her ability to put into words what I wasn't even sure I wanted to say; Iha Yoshiyasu for his early—and ongoing—support; Yoshikawa Hideki, Tobaru Sunao, Robert Avery, Daniel Broudy, Peter Simpson, and the Nakamura-Hubers for their kindness; Kawamura Masami for navigating Japan's freedom of information laws; Ishikawa Bunyo for his lessons on how the Vietnam War traumatized all sides; Kinjo Minoru for his hospitality (and *awamori*); Professor Sakurai Kunitoshi for his explanation of Korea–US environmental agreements; Ina Takahiro and Tamura Susumu for their descriptions of work on the bases; Ampo Yukiko and Jahana Etsuko at the Nuchi du Takara House, Iejima Island; and Uchimura Chihiro at Fukutsukan, Naha City. I would especially like to thank Ginoza Eiko—to list the number of ways you have inspired me would fill a shelf of books; you have taught me the true meaning of *nuchi du takara*.

Since 2010, hundreds of active and retired service members have shared their experiences and advice with me both on and off the record. To those unnamable insiders who risked their careers, compensation, and liberty, you know who you are and you have my heartfelt gratitude. To the families who opened their homes to me and my TV crew in 2012—Larry and Shirley Carlson, John and Angela Santiago, Don and Mary Schneider, Jimmy and Ray Spencer—I owe a massive thanks, particularly Joe Sipala, who showed me most what it means to love your country and fear your government. John Olin and Michelle Gatz, too, conducted extensive research into defoliant use on Okinawa. At the same time, I send an apology to those for whom this book comes too late: L. E.; Larry Gray; Glen Herman; Jerry Mohler; Scott Parton; and Caethe Goetz, who once told me, "There is more power in a multiple of voices than one lone one." Rest in power, and I vow to keep fighting for justice on your behalf.

More than anywhere else in the world the people of Vietnam understand the dangers of military contamination. During my visits there, I was warmly welcomed by all those seeking environmental justice: Nguyen Thi Hien and Phan Thanh Tien of the Da Nang Association of Victims of Agent Orange/Dioxin; Major General Tran Ngoc Tho; and Dr. Nguyen Thi Ngoc Phuong and the staff at Tu Du Hospital, Ho Chi Minh City. Your furious compassion kept me writing, and I hope this book contributes to global understanding of how dioxin impacts your nation's past, present, and future. In addition, I would like to thank Dr. Wayne Dwernychuk for his invaluable input on Vietnam's hot spots and Heather Bowser for sharing her family's hard-learned lessons about dioxin.

On Guam, I am grateful for everyone who took the time to meet with me and answer my questions: senators Sabina Perez and Therese Terlaje,

Attorney General Leevin Camacho, Robert Celestial, the staff of Guam Environmental Protection Agency, and members of the Joint Region Marianas. Ralph Stanton's research on the use of defoliants on Guam was also invaluable.

At Rowman & Littlefield, I'd like to thank those who have worked hard to put this book into print, including Susan McEachern, Katelyn Turner, Alden Perkins, and Nicole Carty. The editor of this Asia/Pacific/Perspectives series, Mark Selden, has been my mentor at the *Asia-Pacific Journal* for the past decade, and I thank him for his constant guidance and friendship. Without Robert Jackson's search of the National Archives, this book would have no cover photo. Thank you.

Near the sea in Naha City stands a memorial to the journalists who died in the Battle of Okinawa, and today many local reporters make an annual pilgrimage there to pledge not to pick up their pens again in the glorification of war. Such consciousness among Okinawan reporters anchors their commitment to fearless, public service journalism. At the *Okinawa Times*, I offer a huge thanks to my colleagues, Yonamine Kazue, Yonahara Yoshihiko, Chinen Kiyoharu, Maeda Takayuki, and Abe Takashi, and at *Ryukyu Shimpo*, Shimabukuro Ryota. Most importantly, I offer my gratitude to *Ryukyu Asahi Housou* TV director Shimabukuro Natsuko for her intelligence, compassion, and eye for a great story; I look forward to working with you for many more years to come.

Also, I'd like to thank the editors of my Japanese books about military contamination, Yamamoto Kunihiko at Koubunken and Nakamoto Naoko at Iwanami Shoten; both publishers remain rare beacons of hope during these dark days.

In Japan—as elsewhere in the world—the media has degenerated from a watchdog to a lapdog for the powers that be, but some survivors remain ready to bare their teeth. I'd like to thank Daimon Sayuri, Eric Johnston, Andrew Kershaw, Elliott Samuels, and Ben Stubbings at the *Japan Times*; Kanda Akira and Okuaki Satoru at NHK; Kanehira Shigenori at TBS; Ota Masakatsu and Sawa Yasuomi at *Kyodo News*; Suzuki Nobuyuki at *Tokyo Shimbun*; Moronaga Yuji at *Asahi Shimbun*; Tsunabuchi Youji at *Shuukan Kinyoubi*; and Greg Starr at the Foreign Correspondents' Club of Japan's *Number One Shimbun*. Also, I appreciate the lessons on environmental contamination I received from Nakamura Goro; Ikeda Komichi; Sakata Masako; Yamauchi Masayuki; and especially Inagaki Mihoko, who opened my eyes to the ongoing impact of Japanese chemical weapons.

Transparency is the cornerstone of a healthy democracy; the US Freedom of Information Act (FOIA) might be flawed, but it still serves an essential role. I thank Beryl Lipton at MuckRock for teaching me its ropes and the FOIA officers whose efforts have helped to illuminate the opacity at the heart of the US–Japan Alliance.

Earlier drafts of this book received invaluable criticism that guided its course, corrected errors, and polished rough corners. I offer enormous thanks to Robert "Bo" Jacobs, Peter Kuznick, LisaLinda Natividad, Douglas Lummis, and Fred Wilcox; I owe you all a drink—or more—and, needless to say, all errors are my own.

Finally, I dedicate this book to my number-one critic, adviser, driver, and friend, Professor Abe Kosuzu, at University of the Ryukyus. *Diolch yn fawr.*

# INDEX

animals: contamination of, 17–18, 25, 42, 84; as early warning system, 74, 78, 192; endangered, 128, 130; habu snakes, 91, 130; invasive species, 178; testing, 11
Anjain, Jeton, 206
Anjain, John, 206
Ansbach base, Germany, 119, 146
anthrax, 2, 14, 21, 23, 24, 27, 228n35
anticorrosion compound, 150, 155, 237n27
aqueous film-forming foam (AFFF), 143–47, *145*
Armed Services Committee (House of Representatives), 66, 159, 178
armor-piercing shells, 106, 134–35
Army Institute of Scientific Research (Japan), 10
arsenic, 15, 17, 19, 59, 97, 133, 215, 249n43; cacodylic acid, 85
*Asahi Shimbun* (newspaper), 171
asbestos, 138, 216
Atlantic Charter, 59
atmospheric testing, ban on, 193
Atomic Bomb Casualty Commission (ABCC), 38, 46
atomic cannons, 67
Atomic Energy Agreement, 164
Atomic Energy Commission (AEC), 46, 49, 194
"The Atomic Plague: I write this as a warning to the world" ( Burchett), 36–37
Atomic Soldiers, 209
atropine (nerve agent antidote), 74, 78
Atsugi Air Base. *See* Naval Air Facility Atsugi (NAFA, Japan)
Australia, 2, 13–14

B-29 Bockscar, 58
B-43 one-megaton bomb, 69, 157
Bainbridge, Kenneth, 32

base commanders, 83, 111, 116, 118–19, 196, 209
benzenes, 104, 124, 133, 141, 216, 236–37n22, 249n43
Bien Hoa Air Base (Vietnam), 89, 120
Bikini Atoll, 4, 41–42, 48–49, 52
biological weapons: anthrax, 2, 14, 21, 23, 24, 27, 228n35; bubonic plague, 23–25, 187, 228n35; cholera, 4, 22, 24; in combat, 12–14, 23–25; human experiments on *maruta*, 22–23; Japanese, 21–23; tests on Okinawa, 72; typhoid, 24
Bionetics Research Laboratories, 86
bioremediation, 190
Blandy, William, 44
blister agent, 78–79
Blue Grass, Kentucky, 197
Bob Hope Primary School (Okinawa), 99, 139
Britain, 2, 9, 82
"Broken Arrows" incident, 69
Brown Agent (hydrogen cyanide), 57
bubonic plague (Black Death), 23–25, 187, 228n35
Burchett, Wilfred, 36–37
Bush, George, 159
"BW activities directed at man" paper, Japanese scientists, 26
Byrnes, James F., 39

cacodylic acid, 85
cadmium, 138, 216–17
Cambodia, 83, 89
Camp Chinen (Okinawa), 61, 189
Camp Courtney (Okinawa), 132–33, 151
Camp Detrick (Fort Detrick, Maryland), 26, 71, 81–82, 92
Camp Edwards, Massachusetts, 246n13
Camp Foster (Okinawa), 60, 151
Camp Hansen (Okinawa), 60, 72, 127–29, 131, 133, 151

# ABOUT THE AUTHOR

**Jon Mitchell** is an investigative journalist with the *Okinawa Times* whose scoops frequently top the TV news and newspapers in Japan. In 2015, he was awarded the Foreign Correspondents' Club of Japan's Freedom of the Press Lifetime Achievement Award for his research on human rights issues on Okinawa. His work has been featured in reports for US Congress and the Japanese parliament; it has also helped US veterans exposed to contamination in Japan win compensation from the Department of Veterans Affairs. US authorities—including the State Department and the Department of Defense—have repeatedly attempted to block Mitchell's work, prompting condemnation from international press freedom groups.

*Defoliated Island*, a TV documentary about Mitchell's investigations into the use of Agent Orange in Japan, was winner of a 2012 award for excellence from the Association of Commercial Broadcasters, a first for a foreign journalist. He is author of two Japanese books about military contamination, *Tsuiseki: Okinawa no Karehazai* (Agent Orange on Okinawa) (2014) and *Tsuiseki: Nichibei Chiikyoutei to Kichi Kougai* (Military Contamination and the Japan–United States Status of Forces Agreement) (2018). In 2019, Okinawa International University unveiled the "Jon Mitchell Collection," a 5,500-page database of reports obtained via the US Freedom of Information Act (FOIA) cataloging the environmental impact of military operations in Japan, CIA attempts to manipulate public opinion, and United States Marine Corps lectures denigrating Okinawans. Mitchell often holds seminars to teach Japanese journalists, academics, and nongovernmental organizations (NGOs) how to use the FOIA.

Mitchell is an *Asia-Pacific Journal* associate and visiting researcher at the International Peace Research Institute of Meiji Gakuin University, Tokyo.

## ASIA/PACIFIC/PERSPECTIVES
## Series Editor: Mark Selden

*Social and Political Change in Revolutionary China: The Taihang Base Area in the War of Resistance to Japan, 1937–1945*
by David S. G. Goodman
*Rice Wars in Colonial Vietnam: The Great Famine and the Viet Minh Road to Power*
by Geoffrey C. Gunn
*Islands of Discontent: Okinawan Responses to Japanese and American Power*
edited by Laura Hein and Mark Selden
*Masculinities in Chinese History*
by Bret Hinsch
*The Rise of Tea Culture in China: The Invention of the Individual*
by Bret Hinsch
*Women in Early Imperial China, Second Edition*
by Bret Hinsch
*Chinese Civil Justice, Past and Present*
by Philip C. C. Huang
*Local Democracy and Development: The Kerala People's Campaign for Decentralized Planning*
by T. M. Thomas Isaac with Richard W. Franke
*Hidden Treasures: Lives of First-Generation Korean Women in Japan*
by Jackie J. Kim with Sonia Ryang
*North Korea: Beyond Charismatic Politics*
by Heonik Kwon and Byung-Ho Chung
*A Century of Change in a Chinese Village: The Crisis of the Countryside*
by Juren Lin, edited and translated by Linda Grove
*Postwar Vietnam: Dynamics of a Transforming Society*
edited by Hy V. Luong
*From Silicon Valley to Shenzhen: Global Production and Work in the IT Industry*
by Boy Lüthje, Stefanie Hürtgen, Peter Pawlicki, and Martina Sproll
*Resistant Islands: Okinawa Confronts Japan and the United States, Second Edition*
by Gavan McCormack and Satoko Oka Norimatsu
*The Indonesian Presidency: The Shift from Personal toward Constitutional Rule*
by Angus McIntyre
*Nationalisms of Japan: Managing and Mystifying Identity*
by Brian J. McVeigh
*Poisoning the Pacific: The US Military's Secret Dumping of Plutonium, Chemical Weapons, and Agent Orange*
by Jon Mitchell
*The Korean War: A Hidden History*
edited by Tessa Morris-Suzuki

*To the Diamond Mountains: A Hundred-Year Journey through
China and Korea*
by Tessa Morris-Suzuki
*To Hell and Back: The Last Train from Hiroshima*
by Charles Pellegrino
*From Underground to Independent: Alternative Film Culture in
Contemporary China*
edited by Paul G. Pickowicz and Yingjin Zhang
*Wife or Worker? Asian Women and Migration*
edited by Nicola Piper and Mina Roces
*Social Movements in India: Poverty, Power, and Politics*
edited by Raka Ray and Mary Fainsod Katzenstein
*Pan-Asianism: A Documentary History, Volume 1, 1850–1920*
edited by Sven Saaler and Christopher W. A. Szpilman
*Pan-Asianism: A Documentary History, Volume 2, 1920–Present*
edited by Sven Saaler and Christopher W. A. Szpilman
*Biology and Revolution in Twentieth-Century China*
by Laurence Schneider
*Contentious Kwangju: The May 18th Uprising in Korea's Past and Present*
edited by Gi-Wook Shin and Kyong Moon Hwang
*Thought Reform and China's Dangerous Classes: Reeducation, Resistance,
and the People*
by Aminda M. Smith
*When the Earth Roars: Lessons from the History of Earthquakes in Japan*
by Gregory Smits
*Subaltern China: Rural Migrants, Media, and Cultural Practices*
by Wanning Sun
*Japan's New Middle Class, Third Edition*
by Ezra F. Vogel with a chapter by Suzanne Hall Vogel,
foreword by William W. Kelly
*The Japanese Family in Transition: From the Professional Housewife Ideal
to the Dilemmas of Choice*
by Suzanne Hall Vogel with Steven K. Vogel
*The Korean War: An International History*
by Wada Haruki
*The United States and China: A History from the Eighteenth Century
to the Present*
by Dong Wang
*The Inside Story of China's High-Tech Industry: Making Silicon Valley
in Beijing*
by Yu Zhou